T0137146

Cognitive Technologies

More information about this series at http://www.springer.com/series/5216

Robert Trappl
Editor

A Construction Manual for Robots' Ethical Systems

Requirements, Methods, Implementations

Editors
Robert Trappl
Austrian Research Institute for Artificial Intelligence (OFAI) of the Austrian Society for Cybernetic Studies (OSGK)
Vienna, Austria

Center for Brain Research,
Medical University of Vienna
Vienna, Austria

Managing Editors
Prof. Dov M. Gabbay
Augustus De Morgan Professor of Logic
Department of Computer Science
King's College London
Strand, London, UK

Prof. Dr. Jörg Siekmann
Forschungsbereich Deduktions- und Multiagentensysteme, DFKI
Saarbrücken, Germany

ISSN 1611-2482 ISSN 2197-6635 (electronic)
Cognitive Technologies
ISBN 978-3-319-79349-8 ISBN 978-3-319-21548-8 (eBook)
DOI 10.1007/978-3-319-21548-8

Springer Cham Heidelberg New York Dordrecht London

Softcover re-print of the Hardcover 1st edition 2015

Printed on acid-free paper

Springer International Publishing AG Switzerland is part of Springer Science+Business Media (www.springer.com)

Preface

In 1906, the Austrian writer Alexander Roda Roda, born Sándor Friedrich Rosenfeld, wrote the story "Das Justizklavier" (The Justice Piano). In some state in the Maghreb—Roda Roda does not locate it precisely—an inventor asks for an audience with the potentate. The potentate, interested in what the man might offer him, receives him. The inventor opens a big trunk and inside there is a small piano. The potentate is curious about this piano, and the inventor explains: "You have big expenses for judges, lawyers, and attorneys, and problems with people who complain about the length of trials. This is my invention, the Justice Piano, that will remove all these problems."

"You see that my piano has white and black keys, like any piano. But these keys are inscribed, the black ones with the possible crimes of the defendant, e.g., 'Burglary', 'Leg Fracture', 'Adultery'. [In German the words rhyme: 'Einbruch', 'Beinbruch', 'Ehebruch'.] The white keys are for extenuating causes, e.g., 'Minor', 'Without Previous Convictions', 'Drunk'. You simply press the white and black keys and without any delay the piano prints out the verdict."

At first, the potentate is enthusiastic but then starts pondering and finally says: "A piano normally has at least two pedals, one for forte and one for piano. Couldn't you attach two pedals, one with the inscription 'Opposition' and the other with 'Government'?"

Things have become far more complex in the meantime, and a simple rule-based expert system would not suffice to help a robot in making ethical decisions. Many papers and even a few books have been published on how to enable robots to make such decisions, a few even claiming that such decisions should not be made by robots. But if it is accepted that robots will have to make decisions which are ethical, then how should a designer of the robot's software proceed? What are the prerequisites of ethical systems, what methods are available, and are there already applications?

In an attempt to answer these questions, the Austrian Research Institute for Artificial Intelligence (OFAI) identified potential contributors, mostly from their previous publications, invited them to submit position papers, and then invited them to a 2-day workshop at the OFAI in Vienna to present and exchange their ideas. The workshop formed the basis for most of the chapters of this book.

The contributors are affiliated with various universities, among them Eindhoven University of Technology; the University of Hartford; Rensselaer Polytechnic Institute; Université Pierre et Marie Curie, Paris; and Universidade Nova de Lisboa.

A book like this would not have been possible without the commitment of many persons. First, I thank the authors who took great pains to enhance their original position papers into book chapters by including new material and by considering the comments in and outside the discussions.

Second, I want to thank my colleagues at the OFAI who have been of great help, especially Karin Vorsteher for her great efforts in proofreading and formatting and for many other activities for which there is not enough space to list.

Third, I thank the series editors, Jörg Siekmann and Dov Gabbay, for including this book in the Cognitive Technologies book series, and the Springer Computer Science editor Ronan Nugent for his support in the publication process.

Finally, I thank the Austrian taxpayers whose money allowed us to develop the workshop, pay for the travel and hotel expenses of the participants, and then prepare this book. We received this money through the Austrian Federal Ministry for Transport, Innovation and Technology, with Doris Bures as then Federal Minister, now President of the Austrian Parliament, and the very supportive officers, Ingolf Schaedler, Michael Wiesmueller, and Karl Supa, to whom I offer my sincere gratitude for trusting that I would finally present a useful product.

I hope you enjoy studying this book.

May 2015 Robert Trappl

Contents

Chapter 1
Robots' Ethical Systems: From Asimov's Laws to Principlism, from Assistive Robots to Self-Driving Cars

Robert Trappl

Abstract This chapter and the book's content should aid you in choosing and implementing an adequate ethical system for your robot in its designated field of activity.

Keywords Ethical system • Assistive robot • Asimov's laws • Principlism • Self-driving cars

1.1 Introduction

In the winter of 2013, the driver of a school bus saw a deer crossing the road and he turned his bus sharply in order not to hit it. The bus skid off the snowy road, rolled down a steep meadow, and was finally stopped by some trunks of trees. Many of the schoolchildren were severely injured and had to be flown to a hospital; it was remarkable that none of them died. The report did not mention the fate of the deer.

Obviously, the driver made an ethical decision, though a wrong one. Many of our decisions are influenced by our ethics, but most of the time we are not aware of this fact. However, we are aware when we decide to act contrary to our moral standards.

When we develop robots to act as, for example, partners in our workplace or as companions when we are old or we have special needs, they need to be equipped with ethical systems for at least two reasons: they should act cooperatively, especially in complex social situations, and they should understand human decisions.

R. Trappl (✉)
Austrian Research Institute for Artificial Intelligence (OFAI), Freyung 6/6, 1010 Vienna, Austria

Center for Brain Research, Medical University of Vienna, Spitalgasse 4, 1090 Vienna, Austria
e-mail: robert.trappl@ofai.at

R. Trappl (ed.), *A Construction Manual for Robots' Ethical Systems*, Cognitive Technologies, DOI 10.1007/978-3-319-21548-8_1

A distinction has to be made between implicit and explicit ethical systems: every robot must, especially in complex social environments, follow ethical principles. However, its ethics can follow implicitly from the decision processes implemented, or its actions should be a consequence of an explicitly designed ethical system in the robot.

It should be stressed that the ethical principles for robots and those for designers, developers, and those who deploy robots need not be identical. This book is concerned with explicit ethical systems for robots.

Furthermore, the ethical system for a robot which is a companion for an older person or a person with special needs will differ from the ethical system needed for a self-driving car: in the first instance, the robot and the human are interacting on a body-to-body basis; in the second case, the human is inside the body of the robot!

While several books recently published give excellent overviews of the research into ethics and robots, e.g., [1–3], this book aims at helping a designer or a developer of robots for a specific purpose to select appropriate ethical rules or an ethical system, and it shows different ways of implementing these.

If they want to do this consistently, we can test the robot's decisions/actions with the comparative moral Turing test, proposed by Allen et al. [4]: an evaluator has to judge decisions made by a human in situations that require ethical decisions and the decisions of the robot in the same situations. If the evaluator cannot determine correctly which made the decision in more than 50 % of the cases, the robot passes the test.

This introductory chapter is divided into three sections: Ethical Systems Usable for Robots; Platforms for Implementation; and Areas for Deployment.

1.2 Ethical Systems Usable for Robots

This section follows partially the descriptions in Anderson and Anderson [5].

The first ethical system proposed for robots appeared in the story "Runaround" from the American author Isaac Asimov, who later, in 1950, included it in a collection of stories in the volume "I, Robot" [6]:

- Law One: A robot may not injure a human being or, through inaction, allow a human being to come to harm.
- Law Two: A robot must obey orders given to it by human beings, except when such orders conflict with Law One.
- Law Three: A robot must protect its own existence as long as such protection does not conflict with Law One or Law Two.

Later, Isaac Asimov added one more law which he named Law Zero, which has to precede Law One (naturally, the "except" phrases had to be changed accordingly):

- Law Zero: A robot may not injure humanity or, through inaction, allow humanity to come to harm.

A second ethical system is based on Jeremy Bentham's "utilitarianism" [7]: its imperative is to act in such a way that the maximum good for all persons involved is obtained. To "act" means to select from all possible actions the appropriate one. However, it is difficult to define "good"; therefore, in most applications it is substituted by "utility." "Utility" is often used by decision or game theorists. In their experiments, life is simplified by using dollars or euros for utilities, thus making utility measurable on a rational scale, at least in some limited range.

Now, in addition to the problems of measuring utilities, there is the problem of calculating the probability with which the person will experience this utility—all of us have probably seen the disappointment of persons we assume we know well when we give them the "wrong" present. Nevertheless, we can risk selecting the optimum action by computing for each potential action the sum of the products of the utility for each person times the probability that each person experiences this utility and then choosing the one with the largest sum.

A third ethical system which may be implemented in robots originates in the realm of medicine. Probably medicine was the first discipline with a professional ethics because the decisions of physicians can have deadly consequences, but also medical ethics seems to be easier to formulate than others. For example, what would be an appropriate ethical system for the actions of lawyers? To win all processes even if you think your client is guilty? To earn the biggest amount of money? Both goals are probably related. Or to accept only poor people as clients? That sounds far more difficult than the case of medicine.

This ethical system, called "principlism" [8], consists of four ethical principles:

1. Autonomy: Respect the autonomy of the person. Not so long ago physicians decided about a therapy, be it conservative or surgery, without asking the patients because they thought they knew better. Today it is impossible, for example, to begin with a surgical intervention without explaining the potential risks to the patient, in detail. In addition, patients have to sign a declaration of informed consent.
2. Beneficence: Your action should bring benefit to the person.
3. Nonmaleficence: Your action should not harm the person. This is a condensed version, in one word, of the Latin commandment "Primum non nocere," in English "Above all, do not do harm."
4. Justice: At first glance a quite surprising principle. However, it is an equally important one: Consider in your action the social (= fair) distribution of benefits and burdens.

Other approaches to ethical systems have been proposed; for both quotations and applications, see, for example, Madl and Franklin [9] or Anderson and Anderson [10].

1.3 Platforms for Implementation

One possible way to implement an ethical system would be to use the Robot Operating System (ROS), originally developed in 2007 by the Stanford Artificial Intelligence Laboratory for the STAIR (STanford AI Robot) project. STAIR's goal description begins with this sentence on its homepage (http://stair.stanford.edu/): "Our single robot platform will integrate methods drawn from all areas of AI, including machine learning, vision, navigation, planning, reasoning, and speech/natural language processing." The question arises where to position the ethical system. One could, for example, argue that its only appropriate place is between lower-level sensors/actuators and any higher-level cognitive system. To see a detailed argumentation, potential severe consequences, and related topics, read Govindarajulu and Bringsjord [11].

Regarding cognitive systems, one example is LIDA (Learning Intelligent Distribution Agent), a cognitive architecture that attempts to model a broad spectrum of cognition in biological systems. It has been mainly developed by Stan Franklin and cooperators [12]; its basis is the Global Workspace Theory developed by Baars [13]. The major processes implemented in the modules of the LIDA cognitive architecture are perception, percept to preconscious buffer, local associations, competition for consciousness, conscious broadcast, recruitment of resources, activation of schemes in the procedural memory, action chosen, and action taken. For a more detailed description of the LIDA architecture and a first attempt to implement moral decision-making in this architecture, see Madl and Franklin [9].

A famous and often used model for many purposes is the BDI (beliefs, desires, intentions) agent model by Rao and Georgeff [14]. The BDI software model implements Bratman's theory of human practical reasoning [15]. This model can be used in the strict sense of the original paper, or its components can be extended to approach human reasoning.

The model has three main components. Beliefs are what you think you know about the outside world. This could be a very restricted subset according to your tasks or, in humans, your acquired knowledge about the world, from the fact that objects can be hard, soft, liquid, etc., to other humans, especially how they feel, think, and act, a "Theory of Mind." They are only "beliefs," as is intuitively obvious when we experience a failure to predict the reaction of a person whom we think we know. Desires are the drives, needs, etc., that should be fulfilled. A simple robot may only desire a fully charged battery, while a robot companion may desire to make a person feel good and even happy for a long period of time. In order to fulfill desires, the agents, both human and robotic, have to ponder how to proceed, given the conditions of the world in their beliefs. Intentions are the results. Intentions can be very simple actions, like finding an electric outlet, to complex strategies requiring careful planning. Probabilities and resulting priorities play important roles. And the resulting actions may be interrupted by "alarms" (e.g., Sloman [16]), for example, when an agent moves an object and suddenly senses that a truck is crossing its path.

An extension of the BDI model incorporating the important area of emotions was undertaken by Rank et al. [17].

A formalization of rational agents within the BDI framework is presented by Ganascia [18], and grafting norms into the BDI agent model by Tufis and Ganascia [19], both in this book. Other approaches for grounding ethics in an agent or a robot are, for example, the use of logic programming in Saptawijaya and Pereira [20] or case-supported principle-based behavior in Anderson and Anderson [10], based on the prima facie duty approach to ethics by Ross [21].

1.4 Areas for Deployment

There already exist so many areas of deployment for robots that this chapter can only focus on two of them: assistive robots and self-driving cars.

An overview of robots as ethical agents, with special consideration of the results of three EU-funded projects that investigated parts of this area, is presented by Krenn [22]. The "Guidelines on Regulating Robotics" (Deliverable D6.2 of the RoboLaw project, supported by the EU; http://www.robolaw.eu) of 2014 also cover ethical aspects of care robots and of self-driving cars.

1.4.1 Assistive Robots

These range from simple vacuum cleaners to robots at the workplace, now no longer behind high fences, and care robots in hospitals to companions for older persons, some of them with special needs. The scenario by Gasser and Steinmüller [23] "Tina and Her Butler" describes how a robot butler could help an older woman to lead a self-determined, independent life in her home. A European Agency, sponsored by the EU, has been founded with the aim of fostering "Active Assisted Living," and such robots are also often called AAL robots. While the obvious ethical system for AAL robots is principlism, other approaches can be successful too, for example, in Anderson and Anderson [10], see the example of the Ethical Eldercare Robot EthEl [24].

An important faculty of assistive robots is speech/language understanding, processing, and expressing. This holds not only for AAL robots which are expected to establish long-term relationships with their clients but also for robots at the workplace. They have to interact with their human co-workers, and, what is more, they should also interact with the other robots in a language understandable to their co-workers, a requirement both for trust-building and for warning their co-workers in the case of emergencies.

Communication is not ethics neutral. Case studies on conversational phenomena show examples of ethical problems on the level of the "mechanics" of conversation, meaning-making, and relationship that have to be considered; see Payr [25].

Another important aspect in assistive robots is their likely influence on the beliefs, opinions, and decisions of humans and their persuasiveness, which also should be considered from an ethical point of view. This topic is investigated in Ham and Spahn [26].

1.4.2 Self-Driving Cars

Recently, a self-driving car drove at 130 km/h (ca. 80 miles/h) on a German Autobahn, with the German Minister for Transportation on board. All the major motor companies, and others, are working to offer self-driving cars as soon as possible. Current forecasts assume a semiautonomous car, i.e., a car that drives autonomously most of the time but needs an alert driver who can rapidly take over control in situations beyond the automatic driving in the next years. A fully self-driving car is expected sometime between 2025 and 2030, but sooner than that would be no real surprise. It is generally agreed that self-driving cars will lead to a drastic reduction in fatalities on roads, an overall reduction in gas consumption, a drastic reduction in the need to construct new roads, and they will enable older persons to be mobile longer than now.

There will be many decisions like the bus driver's dilemma mentioned at the beginning of this chapter, and they will require ethical systems. However, as mentioned above, you are sitting inside this robot! For example, Asimov's Law Two stipulates that the robot should obey the orders of a human as long as it does not conflict with Law One, namely, not to harm a human being. But would you, as a driver, sitting in an SUV, prefer to fall over a cliff in order to avoid crushing a Mini approaching on a narrow road at high speed? Lin [27] discusses several such situations, mentioning, for example, a situation raised by Noah Goodall: an autonomous car is facing an imminent crash. It can select one of two targets to swerve into: either a motorcyclist who is wearing a helmet or a motorcyclist who is not.

If we use as the ethical system either Asimov's laws or utilitarianism or principlism, each would decide to swerve into the motorcyclist wearing the helmet because the risk of harming her/him is lower. But, if such a rule becomes common knowledge, would that not invite motorcyclists not to wear helmets? And if it is known that the ethical systems of self-driving cars will, when faced with a decision between colliding with either a small car, e.g., a Mini, or a big car, e.g., an SUV, decide to crash into the bigger one—will that reduce the sales of big cars? Would a car with such an ethical system become a sales hit? Or should a fair decision depend on the size of the self-driving car, looking for an equal match or should the car driver be enabled to pre-set these decisions on his/her dashboard?

This and the following chapters in this book should aid you in choosing and implementing an adequate ethical system for your robot in its designated field of activity. Today, the topic of ethical systems in self-driving cars raises so many questions that it may form the content of another book.

References

1. Anderson, M., Anderson, S.L. (eds.): Machine Ethics. Cambridge University Press, New York (2011)
2. Lin, P., Abney, K., Bekey, G.A. (eds.): Robot Ethics. The Ethical and Social Implications of Robotics. MIT Press, Cambridge (2012)
3. Wallach, W., Allen, C.: Moral Machines. Teaching Robots Right from Wrong. Oxford University Press, New York (2009)
4. Allen, C., Varner, G., Zinser, J.: Prolegomena to any future artificial moral agent. J. Exp. Theor. Artif. Intell. **12**, 251–261 (2000)
5. Anderson, M., Anderson, S.L.: Machine ethics: creating an ethical intelligent agent. AI Mag. **28**(4), 15–26 (2007)
6. Asimov, I.: I, Robot. Gnome Press, New York (1950)
7. Bentham, J.: An Introduction to the Principles and Morals of Legislation. Clarendon Press, Oxford (1781, reprint 1907)
8. Beauchamp, T.L., Childress, J.F.: Principles of Biomedical Ethics. Oxford University Press, New York (1979)
9. Madl, T., Franklin, S.: Constrained incrementalist moral decision making for a biologically inspired cognitive architecture. In: Trappl, R. (ed.) A Construction Manual for Robots' Ethical Systems: Requirements, Methods, Applications. Springer International Publishing, Cham, Switzerland (2015)
10. Anderson, M., Anderson, S.L.: Case-supported principle-based behavior paradigm. In: Trappl, R. (ed.) A Construction Manual for Robots' Ethical Systems: Requirements, Methods, Applications. Springer International Publishing, Cham, Switzerland (2015)
11. Govindarajulu, N.S., Bringsjord, S.: Ethical regulation of robots must be embedded in their operating systems. In: Trappl, R. (ed.) A Construction Manual for Robots' Ethical Systems: Requirements, Methods, Applications. Springer International Publishing, Cham, Switzerland (2015)
12. Baars, B.J., Franklin, S.: Consciousness is computational: the LIDA model of global workspace theory. Int. J. Mach. Conscious. **1**(01), 23–32 (2009)
13. Baars, B.J.: A Cognitive Theory of Consciousness. Cambridge University Press, Cambridge (1988)
14. Rao, A.S., Georgeff, M.P.: BDI agents: from theory to practice. In: Proceedings of the First International Conference on Multiagent Systems. American Association for Artificial Intelligence (1995)
15. Bratman, M.E.: Intention, Plans, and Practical Reason. CSLI Publications, Stanford (1987)
16. Sloman, A.: How many separately evolved emotional beasties live within us? In: Trappl, R., Petta, P., Payr, S. (eds.) Emotions in Humans and Artifacts. MIT Press, Cambridge (2003)
17. Rank, S., Petta, P., Trappl, R.: Features of emotional planning in software agents. In: Della Riccia, G., Dubois, D., Kruse, R., Lenz, H.-J. (eds.) Decision Theory and Multi-Agent Planning. Springer (2006). Preprint Technical Report: http://www.ofai.at/cgi-bin/tr-online?number+2004-23
18. Ganascia, J.G.: Non-monotonic resolution of conflicts for ethical reasoning. In: Trappl, R. (ed.) A Construction Manual for Robots' Ethical Systems: Requirements, Methods, Applications. Springer International Publishing, Cham, Switzerland (2015)
19. Tufiş, M., Ganascia, J.G.: Grafting norms onto the BDI agent model. In: Trappl, R. (ed.) A Construction Manual for Robots' Ethical Systems: Requirements, Methods, Applications. Springer International Publishing, Cham, Switzerland (2015)
20. Saptawijaya, A., Pereira, L.M.: The potential of logic programming as a computational tool to model morality. In: Trappl, R. (ed.) A Construction Manual for Robots' Ethical Systems: Requirements, Methods, Applications. Springer International Publishing, Cham, Switzerland (2015)
21. Ross, W.D.: The Right and the Good. Oxford University Press, Oxford (1930)

22. Krenn, B.: Robot: multiuse tool and ethical agent. In: Trappl, R. (ed.) A Construction Manual for Robots' Ethical Systems: Requirements, Methods, Applications. Springer International Publishing, Cham, Switzerland (2015)
23. Gasser, R., Steinmüller, K.: Tina and her butler. In: Trappl, R. (ed.) Your Virtual Butler, LNAI 7407. Springer, Berlin (2013)
24. Payr, S., Werner, F., Werner, K.: Potential of Robotics for Ambient Assisted Living. Vienna: FFG benefit. (2015) Available: https://sites.google.com/site/potenziaal/results
25. Payr, S.: Towards human–robot interaction ethics. In: Trappl, R. (ed.) A Construction Manual for Robots' Ethical Systems: Requirements, Methods, Applications. Springer International Publishing, Cham, Switzerland (2015)
26. Ham, J., Spahn, A.: Shall I show you some other shirts too? The psychology and ethics of persuasive robots. In: Trappl, R. (ed.) A Construction Manual for Robots' Ethical Systems: Requirements, Methods, Applications. Springer International Publishing, Cham, Switzerland (2015)
27. Lin, P.: The robot car of tomorrow may just be programmed to hit you. WIRED Magazine, Issue 5 (2014)

Part I
Requirements

Chapter 2
Robot: Multiuse Tool and Ethical Agent

Brigitte Krenn

Abstract In the last decade, research has increasingly focused on robots as autonomous agents that should be capable of adapting to open and changing environments. Developing, building and finally deploying technology of this kind require a broad range of ethical and legal considerations, including aspects regarding the robots' autonomy, their display of human-like communicative and collaborative behaviour, their characteristics of being socio-technical systems designed for the support of people in need, their characteristics of being devices or tools with different grades of technical maturity, the range and reliability of sensor data and the criteria and accuracy guiding sensor data integration, interpretation and subsequent robot actions. Some of the relevant aspects must be regulated by societal and legal discussion; others may be better cared for by conceiving robots as ethically aware agents. All of this must be considered against steadily changing levels of technical maturity of the available system components. To meet this broad range of goals, results are taken up from three recent initiatives discussing the ethics of artificial systems: the EPSRC Principles of Robotics, the policy recommendations from the STOA project *Making Perfect Life* and the MEESTAR instrument. While the EPSRC Principles focus on the tool characteristics of robots from a producer, user and societal/legal point of view, STOA *Making Perfect Life* addresses the pervasiveness, connectedness and increasing imperceptibility of new technology. MEESTAR, in addition, takes an application-centric perspective focusing on assistive systems for people in need.

Keywords Application-centric perspective • Connectedness and increasing imperceptibility of new technology • Ethics for robots as autonomous agents • Human-like communicative and collaborative behaviour • Initiatives discussing the ethics of artificial systems • Pervasiveness • Robots as ethically aware agents • Socio-technical systems • Tool characteristics of robots

---

B. Krenn (✉)
Austrian Research Institute for Artificial Intelligence (OFAI), Freyung 6/6, 1010 Vienna, Austria
e-mail: brigitte.krenn@ofai.at

R. Trappl (ed.), *A Construction Manual for Robots' Ethical Systems*, Cognitive Technologies, DOI 10.1007/978-3-319-21548-8_2

2.1 Introduction

Robots as we knew them in the past were fully controlled technical devices that are either controlled by a computer programme or a human operator. As regards the former, the robots need to operate in closed, non-changing environments, as it is the case for classical industry robots which can, for instance, be found in the automotive, the chemical, the electrical and electronics, the rubber and plastics or the food industries. In classical industry robotics, all possible events and robot actions are known beforehand, and the robot is programmed accordingly. However, there is a strong demand in industry robotics for robots that are flexible enough to easily adapt to new processes and to collaborate in human–robot teams; cf. [1]. Tele-operated robots are a different kind of controlled robots. They can operate in open environments, because human operators interpret the robot's sensory data and steer the robot's actions. These types of robots are typically employed for operation in conditions that are dangerous for humans, such as underwater, in fire incidents and chemical accidents, warfare, and medical operations, e.g. in minimally invasive surgery [2].

In the last decade, research has increasingly focused on the robot as an autonomous agent that knows its goals, interprets sensory data from the environment, makes decisions, acts in accordance with its goals and learns within an action–perception loop. Thus, the robot becomes more apt to autonomously act in open and changing environments. These developments are of interest for both industry and service robotics. Autonomous robots come in different forms. A prominent example in current times are robots as socio-technical systems assisting people in need. The development of robot companions or robot caretakers that support the elderly is of particular interest from a societal point of view. Europe, especially, has to face a growing share of people aged over 65. According to a Eurostat projection from 2013 to 2080, the population aged 65 years or above will account for 28.7 % of the European population (EU-28) by 2080, as compared with 18.2 % in 2013 [3].

Overall, a broad range of aspects must be considered when discussing robot ethics, including the robots' autonomy, their display of human-like communicative and collaborative behaviour, their characteristics of being socio-technical systems designed for the support of people in need, their characteristics of being technical devices or tools with different grades of technical maturity, including the range and reliability of sensor data, the criteria and accuracy guiding their integration and the quality of the thus resulting actions. A different kind of discussion is needed in the context of basic and applied research, regarding the implementation of a policy to create awareness of potential ethical and legal mishaps a certain research or engineering endeavour may lead to and the countermeasures that need to be taken. To be effective, interdisciplinary contexts must be created where technology development will systematically be intertwined with research on ethical (and psychological) impacts of intelligent, life-like artefacts in general and, even more important, in the light of specific application contexts the technology will be developed for. Some of the relevant aspects must be regulated by societal and legal discussion; others may be better cared for by conceiving robots as ethically aware agents. All of this must be considered against steadily changing levels of technical maturity of individual system components.

The chapter is organised as follows: In Sect. 2.2, three recent initiatives/instruments are presented which discuss legal and ethical aspects of intelligent artificial systems from complimentary perspectives, including ethical guidelines for robots as technical devices (Sect. 2.2.1), legal and ethical requirements of human–computer interfaces (Sect. 2.2.2) and guidelines for the ethical assessment of socio-technical applications (Sect. 2.2.3). In the remainder of the chapter, these three perspectives are taken up and applied to a broader discussion of robot ethics, taking into account robots as multiuse tools (Sect. 2.3), the special case of care robots (Sect. 2.4), robot ethics and system functionality (Sect. 2.5). The discussions are concluded in Sect. 2.6.

2.2 Ethics: Setting the Context

The last few years have already demonstrated increased awareness regarding the necessity for regulating legal and ethical issues related to new technologies which act autonomously, which are likely to blur boundaries between life-likeness or human-likeness and technology, and which are used as assistive systems for people in need. Three examples for recent results of discussion are (1) the EPSRC Principles of Robotics (UK, 2011), addressing ethical issues of robots viewed as technical tools rather than autonomous, self-learning systems; (2) the policy recommendations from the STOA project *Making Perfect Life* (EU, 2012), more generally addressing the ethics of "intelligent" computer interfaces; and (3) the MEESTAR model (Germany, 2013) which is an analysis instrument for structuring and guiding the ethical evaluation of socio-technical systems, i.e. systems that interact with and support their human users in everyday life. Whereas each of the initiatives has its specific views on the ethical assessment of such systems, all three taken together support a broader discussion of legal and ethical requirements of socio-technical systems.

2.2.1 EPSRC Principles: Ethical Guidelines for Robots as Multiuse Tools

The UK Engineering and Physical Sciences Research Council published the so-called EPSRC Principles of Robotics in 2011. The principles quoted below are the result of a workshop bringing together researchers from different areas including technology, industry, the arts, law and social sciences. The principles 1 to 5 are quoted from [4].

Principles:

1. Robots are multiuse tools. Robots should not be designed solely or primarily to kill or harm humans, except in the interests of national security.
2. Humans, not robots, are responsible agents. Robots should be designed and operated as far as is practicable to comply with existing laws and fundamental rights and freedoms, including privacy.
3. Robots are products. They should be designed using processes which assure their safety and security.
4. Robots are manufactured artefacts. They should not be designed in a deceptive way to exploit vulnerable users; instead their machine nature should be transparent.
5. The person(s) with legal responsibility for a robot should be attributed.

The EPSRC Principles strongly focus on robots as technical devices, as tools which are used by someone. It is in the nature of tools that they may be used in more than one way and that they are used under human responsibility. For instance, a hammer may be used to nail a picture on the wall, but also to smash somebody's head. How a tool is used is under the responsibility of its user as well as of politics and society providing legal and ethical frames for the uses of the specific kind of tool. In this view on robotic systems, the agent-like aspects of robots, their autonomy, self-learning and adaptive capabilities are not further assessed. However, these are key features of a new generation of robots, which must be addressed, too.

2.2.2 *STOA Project* Making Perfect Life: *Ethical Requirements of Human–Computer Interfaces*

Another workshop, held in 2011, was initiated by the European STOA project *Making Perfect Life*: *Human–Computer Interfaces*. It brought together experts from law, behavioural science, artificial intelligence, computer science, medicine and philosophy. "Implanted Smart Technologies: What Counts as 'Normal' in the 21st Century?" was discussed as overall topic. Results are published in [5]. STOA is the European Parliament's Science and Technology Options Assessment (http://www.europarl.europa.eu/stoa/). The project *Making Perfect Life* (2009–2011) looked into selected fields of engineering artefacts and resulting consequences for policymaking. As for human–computer interfaces, the study distinguishes three types of systems and makes related high-level policy recommendations. Systems are grouped into:

1. **Computers as human-like communication partners**: The computer takes on several roles such as teacher, nurse and friend and acts and communicates accordingly.
2. **Computers as devices for surveillance and alert**: The computer monitors, measures and intervenes with human states such as attention, fatigue, etc.

3. **Ambient intelligence and ubiquitous computing**: The computer becomes more and more imperceptible.

   For details, see [5], p. 130f.
   The resulting policy recommendations address:

1. **Data protection**, specifically for pervasive and highly connected IT systems.
2. **Privacy, transparency and user control** must be embedded in systems design.
3. An **external regulating body** is required to monitor technology developments and issue warnings with respect to ethical, legal and societal challenges.

   See [5], p. 131.

Robots and in particular care robots are realisations of the first two types of systems, i.e. "computers as human-like communication partners" and "computers as devices for surveillance and alert", and they feed data into the third type of systems ("ambient intelligence and ubiquitous computing"), for instance, when they transmit information to applications of telemedicine. They simulate human-like communicative behaviour. They survey and measure their human fosterlings' states, and apart from merely transmitting these data to external services, they are designed to intervene when something goes wrong or moves into an undesirable direction. This immediately leads into ethical discussion of what is (un)desirable under which circumstances, who determines it and according to which criteria. Here the MEESTAR analysis instrument [6] comes into play. It is an attempt to guide the ethical assessment of socio-technical systems. These are systems that interact with and support their human users in everyday life. MEESTAR was developed having in mind assistant systems for the elderly; however, the instrument as such can be applied to assess any socio-technical system.

2.2.3 MEESTAR: Ethical Assessment of Socio-Technical Applications

The major characteristics of MEESTAR are as follows: (1) It is geared to model a specific application scenario, i.e. the specific assistant needs of a concrete person in her/his social context. It does not aim at universal validity. On the contrary, MEESTAR is an instrument to identify at any time the ethical objectionability of concrete applications. The focus lies on identifying and solving ethically problematic effects of the socio-technical system under assessment. (2) The model takes into account the perspectives of different groups of persons including those who use the system such as the elderly, professional caretakers as well as family and friends, the system providers and its developers. A minimal requirement is that the socio-technical system must not do any harm or only a minimum of harm, given the benefit of the system clearly exceeds the harm it may cause. This must be transparent and in consent with the persons concerned.

MEESTAR assessments focus on ethically negative aspects of a socio-technical application. They are guided by questions regarding three levels of analysis:

1. **Ethical dimensions**: care, autonomy/self-determination, safety, equity, privacy, participation and self-conception. At this level, the content of the ethical questions is formulated.
2. **Ethical objectionabilities related to a specific ethical question given a particular application scenario**: A specific socio-technical application may be uncritical, (b) ethically sensitive but can be handled in practic, (c) ethically highly sensitive with need to be constantly monitored, or (d) the application must be rejected because of severe objections.
3. **Perspectives under which 1. and 2. are assessed**: individual, organisational and societal.

While the EPSRC Principles focus on the tool characteristics of robots from a producer, user and societal/legal point of view, STOA *Making Perfect Life* addresses the pervasiveness, connectedness and increasing imperceptibility of new technology. MEESTAR, in addition, takes an application-centric perspective focusing on assistive systems for the elderly. As MEESTAR has been designed for assessing the ethical objectionability of a concrete application for a specific person in her or his social context, the model provides explicit questions for guiding the ethical assessment. MEESTAR assessments are complex qualitative decision processes which cannot be directly implemented on a computer system. However, thinking of robots as autonomous agents with ethical responsibility, the MEESTAR model can be seen as a starting point for deriving capabilities an ethically aware artificial agent should be equipped with. What the EPSRC Principles, the STOA *Making Perfect Life* and MEESTAR can do for developing ethically aware artificial agents will be explored in the following sections.

2.3 Robot Ethics Under the Perspective of Robots as Multiuse Tools

Under the assumption of robots as multiuse tools, the manufacturers and users are responsible for their robots. In this respect, the main discussion in robot ethics must concentrate on the societal and legal frame of robot use. A transparent and broad societal and political discussion of technology is required, in particular of technology which is part of devices which are already in the market or soon to be launched. This is an interdisciplinary endeavour including experts from various fields such as computer science, engineering, AI, ethics, philosophy and law, as well as the general public, especially after expert discussions have reached a certain level of maturity.

In this respect, the formulation of robot ethics requires first of all the articulation of good habits and standards a society and their members should adhere to in the development and use of intelligent, (semi-) autonomous, agentive artificial systems. It is the task of *normative ethics* to devise moral standards that regulate right and wrong conduct; cf. [7].

Understanding a robot as a multifunctional technical device also suggests that robots should be conceived as implicit ethical agents. Therefore, in a first step, we should strive at developing artificial agents whose actions are constrained in such a way that unethical outcome can be avoided. To achieve this, strategies are required to systematically assess the ethical implications of an application, and this is where the MEESTAR framework comes into play. Even though the MEESTAR instrument was developed with focus on caregiving for the elderly, the questions guiding the ethical assessment can be generalised to any socio-technical system. Following is the adapted list of guiding questions. For the original formulation of the questions (in German), see Appendix 1:

1. Is the use of a specific type of assistant system ethically questionable or not?
2. What are the specific ethical challenges?
3. Given the use of a specific kind of assistant systems, is it possible to attenuate or even resolve related ethical problems? If yes, what would be potential solutions?
4. Are there (potential) situations in the use of the system which are ethically so alarming that the system should not be installed and used?
5. Did the use of the system lead to novel and unexpected ethical problems which were not anticipated during the design of the system?
6. What are the specific aspects and functionalities of the system under investigation which require specific ethical care?

Summing up, in a first stage of the development of robot ethics, the following issues must be dealt with:

1. Robots, including sociable robots, are technical devices/multifunctional tools and should be treated as such. This also holds for ethic requirements imposed on robots. Therefore, measures to be taken to implement robot ethics at technology level must accord with the ethical and legal framework devised at societal and political levels. This framework however still needs to be defined.
2. When we talk about robot ethics, we should talk about normative ethics for the use of robots, i.e. right and wrong conduct of robots is the responsibility of the robot users and not of the robots themselves.
3. Following from claim 2, a robot should not be ethical by itself; it should be ethically used. Therefore, robots should be conceived as implicit ethical agents.
4. The discussion about robot ethics should be divided into ethical and legal issues concerning smart and (semi-) autonomous technology (a) that is already integrated or on the verge of being integrated into commercial applications and (b) that is a matter of basic research. While for the former a broad societal consensus and clear legal regulations are required, for the latter, the discussions will be on a more explorative level, together with round tables of groups of experts from various fields, including technology, AI, philosophy, medicine, law, etc.

2.4 The Special Case of Care Robots for the Elderly: Ethical Dimensions Under Assessment in the MEESTAR Model

In Table 2.1, a summary is provided of the ethical dimensions and related questions investigated by MEESTAR, and it is assessed what they mean in terms of intelligent agents. What are the relevant questions for their assessment, and what would be required for their implementation in a robot?

Summing up, the preceding discussion of potential realisations of MEESTAR ethical dimensions within an artificial agent provides input to requirements on modelling mind components for explicit ethical agents.

2.5 Robot Ethics and System Functionality

Robots are a specific type of human–computer interfaces; thus, the considerations from both EPSRC and STOA *Making Perfect Life* hold for robots and determine robot ethical requirements. On the one hand, robots are artefacts, tools and manufactured products for which the human manufacturers and users have legal responsibility. On the other hand, robots are human–computer interfaces that may be designed to simulate human communication and social interaction, to function as devices for surveillance and alert and to operate on data from virtual as well as real-world contexts. They may be equipped with technology that allows them to connect to the internet and to technical devices in their vicinity including smartphones, tablets, sensors and actuators of smart homes. Being computers and hooked up to other computers on which virtually any programme may run, robots do not only have physical presences with specific object/body features but also may create a broad range of virtual presences. This broad potential is constrained by the specific realisation of a particular robot and by its application scenario. Both condition the requirements for the robot to be ethically and legally compliant.

From a point of view of technical realisation, there exists a broad range of mechanisms that may be built into a robot in order to facilitate its ethically compliant use and behaviour. To achieve this, however, we need to know what should constitute ethically compliant behaviour of a specific robot in a concrete application scenario. The definition and formalisation of what is ethical under which conditions are by far harder than their technical implementation. The following is a checklist of technical dimensions that should be considered in order to devise an artificial (implicit or explicit) ethical agent.

Table 2.2 contains a checklist for creating an ethical artificial agent. Guiding questions are posed from a perspective of robots as situated perceptors and actors.

Different constraints for ethical and legal use apply, depending on what can be perceived, which actuators a robot has in use, what the application scenario is and who the users are. Conceiving robots as multifunctional tools also implies the idea of flexible assembly of different functionalities on an individual robot. This

Table 2.1 MEESTAR ethical dimensions, related questions and their potential for realisation in an artificial agent

Ethical dimension	Related questions
Care (Ge.: Fürsorge) To support the ability of a person in need to conduct a self-governed life	Q1: At which point does the technically supported care for a person in need become problematic, because it changes the person's self-esteem and her/his relationship to the world in a way which is not desirable for the person—from her/his own point of view, as well as from an external perspective? Q2: What kind of dependencies in caregiving structures are still acceptable and desirable, and at which point does the positive intention of care turn into paternalism which may be supported or caused by the technical system?
Potential for realisation in an artificial agent/robot	
Needed are intelligent systems that assess, monitor (over time) and foster the user's, i.e. care receiver's, self-esteem and avoid paternalistic behaviours in caregiving. This requires first of all models for the assessment of self-esteem which must be informed by psychology and nursing science and strategies to avoid paternalism in caregiving which also must be informed by nursing science. These models must be implemented in such a way that they are intertwined with the agent's long-term memory LTM and its dialogue system. These are preconditions for making the agent capable of asking questions regarding the assessment of the user's self-esteem and for initiating supportive dialogue	
Autonomy/self-determination(Ge.: Selbstbestimmung) To support freedom of choice and action for the individual To foster social inclusion	Q1: How can people be supported in their right to exercise self-determination? Q2: How can people be supported in their self-determination, for whom the "normal" criteria of self-determined decisions and actions have become questionable or obsolete? Q3: How do we handle the discrepancy that the ascription of self-determination can be in conflict with the demand for care and support?

(continued)

Table 2.1 (continued)

Ethical dimension	Related questions
Potential for realisation in an artificial agent	
Criteria may be implemented to provide levels of choice for people without or with different levels of impairment. This may be achieved by making use of a computer system's capability to constantly monitor its user and environment and to assess the resulting data according to criteria derived from the agent's theory of mind (TOM) of the user and its theory of the user's physical condition. While there is work on modelling TOM in virtual agents and robots [8, 9], computational models of physical condition still need to be developed, taking into account insights from nursing science	
Ethical dimension	Related questions
Safety (Ge: Sicherheit) To prevent the patient to be harmed To ensure immediate service/support in health critical situations To ensure operating safety in the intelligent home To increase objective safety and the subjective feeling of being safe for the person concerned and the caregivers	Q1: How to deal with the effect that the creation of safety (by the socio-technical system) may decrease the existing capabilities of the human (i.e. when people start to rely on technology, they may stop taking care of things themselves)? Q2: How should it be assessed that the assistant system increases the subjective feeling of safety without objectively increasing safety? Q3: How can conflicts be solved between safety and privacy or privacy and self-determination?
Potential for realisation in an artificial agent	
While Q2 and Q3 are subject for ethical discussion at societal level, solutions to Q1 are well suited to be realised as part of the agent's ethical system: A well-funded TOM and theory of the user's physical condition integrated with the agent's LTM and dialogue system allow the agent to encourage the user to do things on her or his own. The artificial agent monitors and assesses the situation and takes initiative only when absolutely needed	

<table>
<tr><th>Ethical dimension</th><th>Related questions</th></tr>
<tr><td>Privacy (Ge.: Privatheit)
On the one hand, age-adequate assistant systems should do their work as discreet and invisible as possible; on the other hand, almost always assistant systems are based on collecting, processing and evaluating sensitive personal data. Both aspects together may be in conflict with the ethically motivated postulation of informed consent</td><td>Q1: How can privacy—above informational self-determination—be assured as a moral right for the individual when designing age-adequate assistant systems?
Q2: How can privacy be protected for cognitively impaired people?
Q3: How to deal with cultural differences in the assessment of the private and the public sphere, e.g. when introducing age-appropriate assistant systems for people with migration background?</td></tr>
<tr><td colspan="2">Potential for realisation in an artificial agent</td></tr>
<tr><td colspan="2">The ethical assessment of privacy comprises the following issues: data protection, protection of privacy in general, protection of privacy for cognitively impaired people as well as cultural differences in what is considered as private and what as public
As stated in the policy recommendations resulting from STOA Making Perfect Life, specific data protection regulations are needed for artificial systems that are increasingly pervasive, distributed and connected. Apart from the necessary societal and legal developments, different levels of data protection and security should be implemented in the respective socio-technical application, so that it is hard to (nearly) impossible for unauthorised persons to access the data collected by the agent. This addresses data stored in the agent's memory as well as data that are transmitted by the agent to external servers.
While data protection and security at the agent level is a matter of low-level technical solution, the protection of the user's privacy lends itself to be modelled as part of the agent's cognitive system, combining LTM, dialogue system, a model of what is considered to be private in the concrete area of application of the given socio-technical system and the TOMs of its users (e.g. the person cared for, the caregivers) augmented with a cultural dimension of discretion. The goal is to steer the agent's dialogue and action strategies. In addition, the agent's culturally augmented user model of the person cared for may also help the caregivers to better understand the individual needs for privacy of the person cared for.
As regards the dialogue capabilities, the agent should be able to articulate which security levels apply to which functionality. Moreover, it should be able to issue warnings in the dialogue with the user, for instance: "Are you sure you want to put this information on Facebook?" or "Did you know that someone from outsight might be able to listen to our conversation?" Equally, the user should be able to tell the agent that some information is strictly confidential and must not be shared with anybody else or only with a certain restricted set of people. Accordingly, the agent needs to be aware to whom it is transmitting what kind of information. As far as communication with other computer systems is concerned, this requires an elaborate concept of data security and its implementation. In addition, it requires the implementation of methods for user identification when it comes to face-to-face communication with humans. This may be done by voice-based speaker identification or other biometric identification methods as face scan, finger or palm print, and iris and retina scan [10–12]</td></tr>
</table>

(continued)

Table 2.1 (continued)

Ethical dimension	Related questions
Equity (Ge.: Gerechtigkeit) To ensure social justice, inter- and intragenerational equity and access to age-appropriate assistant systems	Q1: Who is granted access to age-appropriate assistant systems? Q2: How should age-appropriate assistant systems be financed (how pays how much)? Q3: What is the understanding of intra- and intergenerational justice?
Potential for realisation in an artificial agent	
The ethical dimension equity addresses aspects of socio-technical systems which are outside of the agent and must be regulated by societal and political discussion	
Ethical dimension	Related questions
Participation (Ge.: Teilhabe) To support a self-governed life and equal participation in societal life	Q1: What is the participation of elderly people in societal life, who are not or cannot be part of the labour force anymore? What kind of participation do they wish for themselves? Q2: What kind of participation is (a) aimed at with age-appropriate assistant systems and (b) which one is actually fostered?
Potential realisation in an artificial agent	
Artificial agents have a high potential to foster participation, because of their ability to connect with and monitor the activities in social networks and to influence the group dynamics in virtual communities [13]. The artificial personal assistant may help its user to select appropriate social networks and monitor network activity. In addition, it may support people with special needs in their communication, e.g. in case of typing impairment by making use of intelligent, personalised auto-completion [14], or for vision impaired by making use of text-to-speech technology [15, 16]	

Ethical dimension	Related questions
Self-conception (Ge.: Selbstverständnis)	Q1: How do socio-technical systems account for the question of meaning which may be of particular interest in old age? Q2: In how far changes the tendency of medicalising life cycles the attitude towards age and ageing? Q3: What are the direct or indirect social restraints of dominant views regarding medicalised and technically supported ageing? Q4: In how far establishes age-appropriate technology routines of standardisation?
Potential realisation in an artificial agent	
While questions Q1 to Q3 are a matter of societal discussion, intelligent technology has the potential to counteract routinely grinding-in of treatments and instead support a broader bandwidth of caregiving strategies and behaviours	

Table 2.2 Checklist for creating an ethical artificial agent

<table>
<tr><td>What is the agent's perception space?
What are the agent's perceptors/sensors?</td><td>Does the agent perceive data from digital worlds, from the physical world or from both?
Perceptors can be computer software that collects data from virtual environments (e.g. e-mail, social media, telephone links, queries to search engines, etc.) as well as software that collects data in the physical world (e.g. vision data, audio, biofeedback, etc.)</td></tr>
<tr><td colspan="2">This gives information about which data can be gathered by the system and thus about the necessities for data protection and data security</td></tr>
<tr><td>What is the agent's action space?
What are the agent's actuators?</td><td>Does the agent act in digital and/or physical environments?
Actuators can be computer software that triggers individual actions in the agent's virtual or physical environments. In this context, it is important to assess what is the (potential) outcome of each single action the agent is capable of</td></tr>
<tr><td colspan="2">These considerations help to assess the potential of each agent action to do harm in the virtual and/or physical world. Becoming clear about this is a precondition to devise respective control mechanisms, internal and external to the agent</td></tr>
<tr><td>What are the dimensions of autonomy built into the agent?</td><td>This question mainly addresses the working mechanisms of the agent's interpretation and control layers, i.e. those aspects of the system that interpret the sensor data and decide upon which actions will be triggered. This leads to further questions, including: Who is the actor and who has control over the action—the human, the robot or both? To which extent and related to which aspects are learning mechanisms employed in the system components?</td></tr>
<tr><td colspan="2">With increasing autonomy of the system, agenthood is shifted from the human to the robot, and the more it is necessary to build mechanisms into the artificial agent that allow the agent itself to be aware of its actions and their potential effects and to be transparent about the reasons for the respective actions</td></tr>
<tr><td>What is the degree of human-likeness in the agent's appearance and behaviour?</td><td>Does the system aim at an illusion of human-likeness, e.g. does it engage in natural language communication, to which extent has the agent features of a human body, does the agent emit socio-emotional signals and does it engage in social interaction?</td></tr>
<tr><td colspan="2">These questions help to assess how likely it is that the agent's simulation of human-likeness will deceive its human user. Depending on the application scenario, the user's knowledge about the technology and her or his mental condition, the assessment of one and the same technical device may lead to different results</td></tr>
</table>

requires certain mechanisms that allow for flexible connection and disconnection of functionalities on the robot at perception and action levels as well as their integration into the robot's control mechanisms (mind), also including mechanisms that support ethically compliant robot action. This requires:

- An action–perception architecture that allows to connect and disconnect action and perception components, i.e. the agent's tools and senses to interact with the outside world, be it a virtual or a physical one
- Models and mechanisms to structure the agent's knowledge of self, others and the environment it is acting in
- Mechanisms that generate natural language utterances based on the agent's memory content and its various models of self, others and the environment

For initial work in this direction, see, for instance, [17–19].

2.6 Conclusion

The formulation of ethical principles for robots has different facets and is a moving target, especially as the technical developments in modelling self-learning, autonomy and natural language faculty are successively improving. Depending on the technical realisation of a robot and its area of application, different requirements regarding robot ethics apply, including question of legal liability, data collection and privacy as well as the rights of those people who are given care by assistive robots. In this chapter, three recent initiatives debating aspects of the above-mentioned requirements are discussed, including the EPSRC Principles of Robot Ethics, the STOA project *Making Perfect Life*: Human–Computer Interfaces and the MEESTAR instrument for assessing the ethical implications of socio-technical systems. While the EPSRC Principles focus on robots as multifunctional technical devices their human producers and users are responsible and liable for, the STOA project *Making Perfect Life* defines policy recommendations for computer systems that act as human-like communication partners and surveillants, and the MEESTAR model is devised to guide the ethical assessment of socio-technical systems in concrete application scenarios.

Understanding a robot as a multifunctional technical device also suggests that the robot should be conceived as implicit ethical agent. In this respect, it is argued in the chapter that, first of all, developers should strive at creating artificial agents whose actions are constrained in such a way that unethical outcome can be avoided. In this respect, creating an implicit ethical agent is an issue of robot design. To find out about relevant design criteria, strategies are required to systematically assess the ethical implications of concrete applications, and MEESTAR provides a framework for this kind of assessment. Furthermore, in this chapter, the MEESTAR ethical dimensions and related questions are assessed with respect to their potential for realisation in an artificial agent's mind. For instance, while data protection and security at agent level is a matter of low-level technical solution suitable to be realised in an implicit ethical agent, the protection of the user's privacy lends itself to be modelled as part of the agent's cognitive system, combining long-term memory, dialogue system, a model of what is considered to be private in a concrete area of application of a given socio-technical system and respective theories of mind (TOM) of the agent's users (e.g. the person cared for and the caregivers) augmented with a cultural dimension of discretion. This already requires the realisation of explicit ethical agents capable of identifying and interpreting relevant information and deriving ethically sound behaviours. For a distinction of implicit and explicit ethical agents, see, for instance, [20].

Overall, two bodies of questions arise for the development of ethically aware agents: (1) How to determine what we expect from an ethical agent? This includes questions such as: In which sense an artificial agent should be ethical? What are the ethical requirements we pose on robots in specific application scenarios? How do we determine these requirements? Instruments such as MEESTAR help to further assess these questions. (2) What are the preconditions to be modelled and technically implemented in order to create ethically aware artificial agents? This implies questions such as: What kind of ethically aware artificial agent can be realised given the state-of-the-art in technical as well as in model development? For instance, well-funded TOM models and theories of users' mental and physical condition are required for health care and assistant robots. Accordingly, developing ethical agents not only requires close collaboration between technicians such as computer scientists and AI researchers, philosophers and lawyers but also must include experts from the specific application domains an artificial agent is going to be developed/deployed for.

Appendix 1: MEESTAR Guiding Questions Original Formulation (German)

1. Ist der Einsatz eines bestimmten altersgerechten Assistenzsystems ethisch bedenklich oder unbedenklich?
2. Welche spezifisch ethischen Herausforderungen ergeben sich durch den Einsatz eines oder mehrerer altersgerechter Assistenzsysteme?
3. Lassen sich ethische Probleme, die sich beim Einsatz von altersgerechten Assistenzsystemen ergeben, abmildern oder gar ganz auflösen? Wenn ja, wie sehen potenzielle Lösungsansätze aus?
4. Gibt es bestimmte Momente beim Einsatz eines altersgerechten Assistenzsystems, die ethisch so bedenklich sind, dass das ganze System nicht installiert und genutzt werden sollte?
5. Haben sich bei der Nutzung des Systems neue, unerwartete ethische Problempunkte ergeben, die vorher – bei der Planung oder Konzeption des Systems – noch nicht absehbar waren?

6. Auf welche Aspekte und Funktionalitäten des untersuchten altersgerechten Assistenzsystems muss aus ethischer Sicht besonders geachtet werden?

   Quoted from [6], p. 14.

Appendix 2: Ethical Dimensions Assessed in MEESTAR, Original Formulation (German)

All quotes [6], pp. 16–20.

Ethical dimension	Related questions
Care (Ge.: Fürsorge)	Q1: "An welchem Punkt wird eine technisch unterstützte Sorge für hilfebedürftige Menschen problematisch, weil sie das Selbstverhältnis und das Weltverhältnis dieser Menschen auf eine Weise verändert, die diese selbst nicht wünschen bzw. die wir Anderen im Blick auf diese Menschen nicht wünschen sollen?" p. 16 Q2: "Welche Grade der Abhängigkeit in Fürsorgestrukturen sind noch akzeptabel bzw. gewünscht und ab welchem Punkt wird aus positiv gemeinter Fürsorgehaltung eine Bevormundung bzw. eine negativ bewertete paternalistische Einstellung, die unter Umständen technisch unterstützt bzw. hergestellt werden kann?" p. 16
Autonomy/self-determination (Ge.: Selbstbestimmung)	Q1: "Wie können – in Anlehnung an eine konsequent am Selbstbestimmungsrecht des Einzelnen orientierte Praxis – Menschen bei der Ausübung ihrer Selbstbestimmung unterstützt werden?" p. 16 Q2: "Wie können Menschen in ihrer Selbstbestimmung unterstützt werden, bei denen die 'normalen' Kriterien selbstbestimmten Entscheidens und Handelns fraglich oder gar hinfällig geworden sind?" p. 16 Q3: "Wie gehen wir damit um, dass die Zuschreibung von Selbstbestimmung mit dem Anspruch auf Fürsorge und Unterstützung in Konflikt treten kann?" p. 16
Safety (Ge: Sicherheit)	Q1: "Wie ist dem zu begegnen, das die Herstellung von Sicherheit unter Umständen zur Verringerung vorhandener Fähigkeiten führt, d.h. wenn Menschen beginnen, sich auf Technik zu verlassen, hören sie vielleicht auf, sich selbst um bestimmte Dinge – in einem produktiven Sinn – zu sorgen?" p. 17 Q2: "Wie ist es zu bewerten, wenn durch ein Assistenzsystem das subjektive Sicherheitsgefühl steigt, ohne dass objektiv die Sicherheit erhöht wurde?" p. 17 Q3: "Wie können Konflikte zwischen Sicherheit und Privatheit oder Sicherheit und Selbstbestimmung (Freiheit) gelöst werden?" p. 17

(continued)

Ethical dimension	Related questions
Privacy (Ge.: Privatheit)	Q1: "Wie kann die Privatsphäre des Einzelnen über die informationelle Selbstbestimmung hinaus als moralischer Anspruch bei der Gestaltung altersgerechter Assistenzsysteme zur Geltung gebracht werden?" p. 18 Q2: "Wie kann die Privatheit kognitiv eingeschränkter Menschen geschützt werden?" p. 18 Q3: "Wie ist mit kulturellen Unterschieden in der Bewertung von privater und öffentlicher Sphäre umzugehen – z.B. bei Einführung von altersgerechten Assistenzsystemen bei Menschen mit Migrationshintergrund?" p. 18
Equity (Ge.: Gerechtigkeit)	Q1: "Wer bekommt Zugang zu altersgerechten Assistenzsystemen?" p. 18 Q2: "Wie soll die Finanzierung von altersgerechten Assistenzsystemen gestaltet werden (wer zahlt wie viel)?" p. 18 Q3: "Welches Verständnis von intragenerationeller und intergenerationeller Gerechtigkeit liegt vor?" p. 18
Participation (Ge.: Teilhabe)	Q1: "Welche Teilhabe besteht für ältere Menschen, die nicht mehr in das Arbeitsleben integriert werden (sollen)? Welche Teilhabe wünschen sie sich?" p. 18 Q2: "Welche Art und Weise der Teilhabe wird durch altersgerechte Assistenzsysteme a) anvisiert und b) tatsächlich gefördert? Inwiefern werden durch technische Assistenzsysteme bestimmte Teilhabevarianten be- oder verhindert?" p. 18
Self-conception (Ge.: Selbstverständnis)	Q1: "Wie wird der Sinnfrage, die im Alter verstärkt auftreten mag, Raum und Perspektive in sozio-technischen Arrangements geboten?" p. 19f Q2: "Inwiefern verändert die Tendenz zur Medikalisierung des Lebens auch die Haltung zum Alter und Altern?" p. 19f Q3: "Welche (direkten oder auch indirekten) sozialen Zwänge entstehen durch dominante Bilder des medikalisierten bzw. technisch unterstützten Alter(n)s?" p. 19f Q4: "Inwiefern werden durch altersgerechte Technik Normierungsroutinen etabliert?" p. 19f

References

1. World Robotics – Industrial Robots 2014. Statistics, Market Analysis, Forecasts and Case Studies. IFR Statistical Department, Frankfurt, Germany. http://www.worldrobotics.org/uploads/media/Executive_Summary_WR_2014_02.pdf http://www.worldrobotics.org/uploads/media/Foreword_2014_01.pdf
2. Lichiardopol, S.: A Survey on Teleoperation. DCT Report, Technische Universiteit Eindhoven (2007)
3. Eurostat, http://ec.europa.eu/eurostat/statistics-explained/index.php/Population_structure_and_ageing
4. EPSRC Principles of Robotics, http://www.epsrc.ac.uk/research/ourportfolio/themes/engineering/activities/principlesofrobotics/
5. van Est, R., Stemerding, D. (eds.): Making Perfect Life. European Governance Challenges in 21st Century Bio-engineering. European Parliament STOA – Science and Technology Options Assessment (2012) http://www.rathenau.nl/uploads/tx_tferathenau/Making_Perfect_Life_Final_Report_2012_01.pdf

6. Manzeschke, A., Weber, K., Rother, E., Fangerau, H.: Studie “Ethische Fragen im Bereich Altersgerechter Assistenzsysteme”. VDI/VDE Innovation + Technik GmbH, Januar (2013)
7. Internet Encyclopedia of Philosophy, http://www.iep.utm.edu/ethics/
8. Hiatt, L.M., Harrison, A.M., Trafton, G.J.: Accommodating human variability in human–robot teams through theory of mind. In: Walsh, T. (ed.) Proceedings of the Twenty-Second International Joint Conference on Artificial Intelligence (IJCAI’11), vol. 3, pp. 2066–2071. AAAI Press (2011)
9. Pynadath, D.V., Si, M., Marsella, S.C.: Modeling theory of mind and cognitive appraisal with decision-theoretic agents. In: Gratch, J., Marsella, S. (eds.) Social Emotions in Nature and Artifact, pp. 70–87. Oxford University Press, Oxford (2013)
10. Orság, F.: Speaker recognition in the biometric security system. Comput. Inform. **25**, 369–391 (2006)
11. Chakraborty, S., Bhattacharya, I., Chatterjee, A.: A palmprint based biometric authentication system using dual tree complex wavelet transform. Measurement **46**(10), 4179–4188 (2013)
12. Dehghani, A., Ghassabi, Z., Moghddam, H.A., Moin, M.S.: Human recognition based on retinal images and using new similarity function. EURASIP J. Image Video Process. **2013**, 58 (2013)
13. Skowron, M., Rank, S.: Interacting with collective emotions in e-Communities. In: Von Scheve, C., Salmela, M. (eds.) Collective Emotions, Perspectives from Psychology, Philosophy, and Sociology. Oxford University Press, Oxford (2014)
14. Matiasek, J., Baroni, M., Trost, H.: FASTY – a multi-lingual approach to text prediction. In: Miesenberger, K., et al. (eds.) Computers Helping People with Special Needs, pp. 243–250. Springer, Heidelberg (2002)
15. King, S., Karaiskos, V.: The blizzard challenge 2009. In: Proceedings of the International Blizzard Challenge TTS Workshop (2009)
16. Laghari, K., et al.: Auditory BCIs for visually impaired users: should developers worry about the quality of text-to-speech readers? In: International BCI Meeting 3–7 June, Pacific Grove, CA (2013)
17. Eis, C., Skowron, M., Krenn, B.: Virtual agent modeling in the RASCALLI platform. In: PerMIS’08 – Performance Metrics for Intelligent Systems Workshop, August 19–21, pp. 70–76. Gaithersburg, MD, USA (2008)
18. Skowron, M., Irran, J., Krenn, B.: Computational framework for and the realization of cognitive agents providing intelligent assistance capabilities. In: The 18th European Conference on Artificial Intelligence Proceedings 6th International Cognitive Robotics Workshop, July 21–22, pp. 88–96. Patras, Greece (2008)
19. Gregor Sieber, G., Krenn, B.: Towards an episodic memory for companion dialogue. In: Allbeck, J., et al. (eds.) Intelligent Virtual Agents, LNAI 6356, pp. 322–328. Springer, Heidelberg (2010)
20. Moor, J.H.: The nature, importance, and difficulty of machine ethics. In: Anderson, M., Anderson, S.L. (eds.) Machine Ethics, pp. 13–20. Cambridge University Press, New York (2011)

Chapter 3
Towards Human–Robot Interaction Ethics

Sabine Payr

Abstract Social assistive robots are envisioned as supporting their users not only physically but also by communicating with them. Monitoring medication, reminders, etc., are typical examples of such tasks. This kind of assistance presupposes that such a robot is able to interact socially with a human. The issue that is discussed in this chapter is whether human–robot social interaction raises ethical questions that have to be dealt with by the robot. A tour d'horizon of possibly related fields of communication ethics allows to outline the distinctive features and requirements of such an "interaction ethics". Case studies on conversational phenomena show examples of ethical problems on the levels of the "mechanics" of conversation, meaning-making, and relationship. Finally, the chapter outlines the possible connections between decision ethics and interaction ethics in a robot's behaviour control system.

Keywords Human–robot–interaction • Social interaction • Interaction ethics • Machine ethics

3.1 About Robots and Social Interaction

3.1.1 What Kind of Robot?

In this chapter, I will be concerned with social assistive robots. This class of robots could easily be confused with what [1] defined as socially assistive robots. While assistive robotics (AR) has largely referred to robots that assisted people with physical disabilities through physical interaction, this definition did not cover assistive robots that assist through non-contact interaction, such as those that interact with convalescent patients in a hospital or senior citizens in a nursing home. The term socially interactive robotics (SIR) was first used by [2] to describe robots whose main task was some form of interaction. The term was introduced to distinguish social interaction from teleoperation in human–robot interaction (HRI).

S. Payr (✉)
Austrian Research Institute for Artificial Intelligence (OFAI), Freyung 6/6, 1010 Vienna, Austria
e-mail: sabine.payr@ofai.at

R. Trappl (ed.), *A Construction Manual for Robots' Ethical Systems*, Cognitive Technologies, DOI 10.1007/978-3-319-21548-8_3

Fong et al. [2] conducted a survey of socially interactive robots and evaluated them along social interaction principles, categorising them by the aspects of social interaction (speech, gestures, etc.) they used. Concerns regarding human perception of robotics, particularly the difference in social sophistication between humans and social robots, were addressed, and field studies, evaluation, and long-term interaction were all noted as areas worthy of future research.

In [1], socially assistive robotics (SAR) shares with assistive robotics the goal to provide assistance to human users, but it specifies that the assistance is through social interaction. Because of the emphasis on social interaction, SAR has a similar focus as SIR. But while a socially interactive robot's goal is to develop close and effective interactions with the human for the sake of interaction itself, in SAR, the robot's goal is to create close and effective interaction with a human user for the purpose of giving assistance and achieving measurable progress in convalescence, rehabilitation, learning, etc.

Their reason for excluding any physical assistance in this definition is not entirely clear. It may have been the reason why this segmentation of the field did not catch on. In the recently published standard ISO 13482:2014,[1] one can find another attempt at classifying what is called here personal care robots and jointly defined by assistive actions to improve the quality of life, including actions where physical contact with the human takes place. Here we find:

- Mobile servant robots, doing tasks in cooperation with the human, including manipulation of objects and exchange of information (e.g. companion robots)
- Physical assistant robots which compensate or support physical abilities (e.g. exoskeletons)
- Person carrier robots which transport people (e.g. robotic wheelchairs)

The standard explicitly excludes robots for medical and military use, industrial and toy robots, robots moving faster than 20 km/h (i.e. robotic cars) as well as swimming and flying robots. What is missing, however, is the mention of tasks like physical and cognitive training, rehabilitation, and health/safety monitoring of patients which Feil-Seifer and Mataric [1] had focused on in their definition. Tasks like these may be considered as implicitly included in the mobile servant's functions under the heading of "exchange of information" (if we allow for this persistent reductionist view of social interaction), but the classification would exclude, say, a socially interactive robotic wheelchair. I therefore chose the term "social assistive robots" to include all those personal care robots that are capable of social interaction, regardless of their physical appearance or other functionality, but (in contrast to [1]) including physical contact with humans.

---

[1] ISO 13482:2014 Robots and robotic devices—Safety requirements for personal care robots. http://www.iso.org/iso/catalogue_detail?csnumber=53820.

3.1.2 Can There Be Social Interaction with Machines?

In sociology, social interaction is a dynamic sequence of social actions between individuals (or groups) who modify their actions and reactions due to actions by their interaction partner(s). Social acts, for their part, are the acts, actions or practices of two or more people mutually oriented towards each other's selves, that is, any behaviour that tries to affect or take account of each other's subjective experiences or intentions. This means that the parties to the social interaction must be aware of each other. Wherever people treat each other as object, things or animals or consider each other as reflex machines or only cause–effect phenomena, there is no social interaction [3]. In other words, if a cyclist knocks down a pedestrian, this is not a social interaction, but the argument they have afterwards is one.

The definition comes relatively easy as long as we limit ourselves to humans as the acting parties in social interaction, but if one of the agents is a robot, we need to revise it. For some, the fact that a robot is capable of autonomous action makes it an agent, but not necessarily a social one. One could discuss endlessly whether a robot that recognises speech and can respond to it by uttering suitable scripted sentences is such a social agent. Such a variation on the Chinese Room [4], ruminating the question whether there is such a thing as an "essential social-ness", would not take us any further. Instead, an argument from sociology helps us there: even if I interact with another human, I can only assume that this being is a social agent, i.e. is aware of me as a human being. As long as the other behaves more or less normally (i.e. within social norms), there is no reason to assume anything else or even to test my assumption. This experience of social normality leads humans to extend this assumption to robots, in a sense. It has been shown that they test robots, e.g. with questions of which they assume that only humans can answer. The curious observation that can be made is, however, that in their interaction behaviour they remain "social" in most of the cases [5]. They tend to make conversational openings and closings, they take orderly turns, and they continue arguing instead of pulling the plug. The answer whether robots can interact socially has been given empirically by humans who act as if they could [6]. This does not mean that humans need to believe naively that the robot has life, conscience, intelligence, emotions, or whatever human-like feature: humans are not as simple as that. If asked, they would have no problem—at least with today's robots—to see them as machines. But they are ready to and, it seems, even drawn into, a kind of game in which they genuinely act as if the robot were a social agent and, as such, interacting socially.[2]

The curious answer to the question whether robots are capable of social interaction, then, is that they are capable of what humans assume they are. Obviously, one social agent who assumes the other to be a social agent, too, seems to be enough to make their interaction social, and robots, call systems, virtual characters, and the like exploit this human disposition towards sociality.

[2]This double game appears less astonishing when one considers how children play with their dolls and stuffed animals: one minute they are cuddled and "alive", the next minute a neglected object, with an apparently effortless switch between the two modes.

One can make an ethical issue out of this deception and put into question the whole idea of having machines use any communication channel that motivates such assumptions and expectations in humans. Although this is a legitimate question, it is not within the scope of this book. For what follows, we will take as granted the possibility of social interaction between humans and robots and go on from there to ask what ethical issues have to be considered within this conceptual framework. More precisely, we will study what ethical questions are raised by social interaction as such, i.e. not only by what information is exchanged, but also by how this is done with regard to the other as a social being.

To place this chapter conceptually inside this book, I will use the case of the eldercare robot EthEl designed by Anderson and Anderson (this volume, [7]) on the basis of a prima facie duty approach. The scenario is the following: the robot has to remind a patient to take a medication and notify an overseer when the patient does not comply. The robot must balance three duties: ensuring that the patient receives a possible benefit from taking the medication, preventing the harm that might result from not taking the medication, and respecting the autonomy of the (adult and competent) patient [8]. Provided with input including the time for medication, the amount of harm that could be done by not taking it, the amount of good to be derived from taking it, and the amount of time within which the benefit would be lost, the robot then determines the change in duty satisfaction and violation levels over time and takes ethically informed decisions about when to remind the patient or when to call in the overseer.

Compared with the ethicists' all-time favourite stories involving dragons, trolleys, or drowning sailors, this scenario is noticeably more realistic. As soon as we consider a realistic scenario, and specifically such a profoundly social one as that of the eldercare robot, we realise that the decisions have to be carried out in social actions and interactions. Reminders and notifications could be done in a hundred different ways, and the hypothesis with which we start into the discussion of a human–robot interaction ethics is that these ways are not irrelevant from the ethical point of view.

The rest of this chapter is organised as follows: First, we will look into potentially related fields of (applied) ethics to find a basis for describing the domain of an ethics of social interaction. We will then look more closely into the layers and aspects in which questions of such an ethics of social interaction could arise, addressing the mechanisms, the contents, and the relational aspects respectively. We will then be able to discuss the role of an interaction ethics for robots both with regard to an ethical and a social–relational system for a robot.

3.2 Related Fields of Ethics

In this section, we will examine whether ethical questions in human–robot interaction can be described and dealt with in existing conceptual frameworks. Obviously, communication ethics comes to mind first as a related domain, with several subdomains such as media communication, organisational communication, and interpersonal/dialogic ethics.

3.2.1 Ethics of the Communication Industry

Communication ethics is, as defined by the international institute of this name,[3] defines the field as "a discipline that supports communication practitioners by offering tools and analyses for the understanding of ethical issues". From its aims, it is clear that communication ethics is understood here as the ethics of the communication industries and their professionals. Although contradicting the very definition of "communication" as an exchange, it is defined here primarily as a one-way distribution of messages and focuses on message producers and senders, and so do the ethical considerations.

The National Communication Association (NCA)[4] adopts the transactional model and so comes closer to a two-way view of communication. The transactional model of communication [9] stresses the reciprocity of the process and represents the collaborative and ongoing message exchange between individuals, or an individual and a group of individuals, with the goal of understanding each other (Fig. 3.1).

Consequently, we would expect the Credo of ethics developed by the NCA [10] to include both sender and receiver of the message and the exchange aspect of communication, but even if some principles may also touch upon the receiver, the focus remains clearly on the producers and senders of messages.

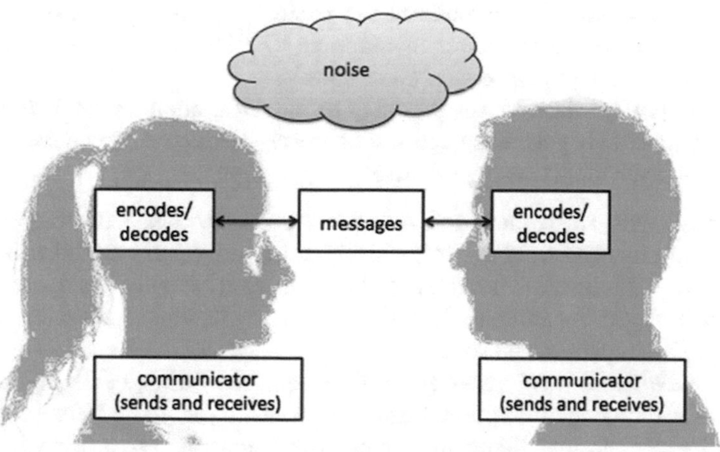

Fig. 3.1 The transactional model of communication (after NCA)

[3]Institute of Communication Ethics, http://www.communicationethics.net/home/index.php.

[4]https://www.natcom.org/about/.

3.2.2 Ethics of Organisational Communication

Literature in organisational communication seems to take the receiver more into account. Andersen [11] explains that "It [ethics] is a dimension that is relevant to all the actors in the communication process—the source or the originator, the person that initiates communication; the person who receives, interprets, hears, reads the communication; and the people who, in effect, are further agents of transmission". In essence, Andersen sees communication ethics as something that needs to be examined from both the source and receiver's point of view, but he also realises that understanding ethics from a societal viewpoint is important.

The source's ethical choices involve her or his basic intent towards her or his receiver(s) [12]. In other words, when determining whether a specific communicative interaction was ethical from a source's perspective, the goodness of the source's intent is what should be examined instead of examining the message itself.

The receiver's ethical choices involve how the individual decides to process the message being sent by the source. The idea of a receiver ethic starts with the notion that being a receiver of a message should be an active process and not a passive process. The receiver has a responsibility to listen, to be critical, to evaluate, to reject, to demand more information, and to reject, whatever the case may be. However, there is another aspect to receiver ethics that must also be considered. As noted by [13], receivers must attend to a message objectively. Quite often receivers attend to messages depending on either their initial perception of the message or their initial perception of the sender. When these initial perceptions interfere with our ability as a receiver to listen, be critical, evaluate, or reject a message, we are not ethically attending to a message.

Overall, Andersen [11] summarises his position by stating, "So, one begins to say that in all the activity of communication, in whatever role we may happen to be in at the moment, there is an ethical dimension".

Redding [14] presents a basic typology of unethical organisational communication consisting of six general categories: coercive, destructive, deceptive, intrusive, secretive, and manipulative–exploitative.

- **Coercive acts** are communication events or behaviours reflecting abuses of power or authority resulting in (or designed to effect) unjustified invasions of autonomy. This includes intolerance of dissent, restrictions of freedom of speech, refusal to listen, resorting to formal rules and regulations to stifle discussion or to squash complaints, and so on.
- **Destructive acts** attack receivers' self-esteem, reputation, or deeply held feelings; reflecting indifference towards, or content for, basic values of others. This includes insults, derogatory innuendoes, epithets, jokes (especially those based on gender, race, sex, religion, or ethnicity), put-downs, back-stabbing, character assassination, and so on. It also includes the use of "truth" as a weapon (as in revealing confidential information to unauthorised persons or in using alleged "openness" as a façade to conceal the launching of personal attacks. It also can include silence: failure to provide expected feedback (especially

recognition of good work), so that message senders (e.g. managers) are perceived as being cold, impersonal, unfeeling, self-centred, and so on. When looking at Redding's explanation of destructive communicative acts, there are clearly two parts: aggressive communication and use of information.

- **Deceptive acts** are communication events or behaviour reflecting a wilful perversion of the truth in order to deceive, cheat, or defraud. This includes evasive or deliberately misleading messages, which in turn includes equivocation (i.e. the deliberate use of ambiguity), also bureaucratic-style euphemisms designed to cover up defects, to conceal embarrassing deeds, or to "prettify" unpleasant facts. In this category of unethical behaviour, we have non-truthful and misleading messages.
- **Intrusive acts** are communication behaviour that is characteristically initiated by message receivers. For example, the use of hidden cameras, the tapping of telephones, and the application of computer technologies to the monitoring of employee behaviour, in other words, surveillance. The fundamental issue, of course, revolves around the meaning and legitimacy of "privacy rights".
- **Secretive acts** are various forms of nonverbal communication, especially (of course) silence and including unresponsiveness. It includes such behaviours as hoarding information ("culpable silence") and sweeping under the rug information that, if revealed, would expose wrongdoing or ineptness on the part of individuals in positions of power.
- **Manipulative–exploitative acts**: [14] defined manipulative–exploitative acts as those where the source purposefully prevents the receiver from knowing the source's actual intentions behind a communicative message. A term that Redding finds closely related to these unethical acts is demagoguery: a demagogue is one who, without concern for the best interests of the audience, seeks to gain compliance by exploiting people's fears, prejudices, or areas of ignorance. Closely related to, if not a variant of, demagoguery is the utterance of messages that reflect a patronising or condescending attitude towards the audience—an unstated assumption that audience members are dull-witted, immature, or both.

The receiver's role is made clear in some of these types of unethical acts. What should be noted is that a large part of the receiver's unethical behaviour consists not in acts, but in "non-acts": silence, unresponsiveness, withholding of feedback, non-disclosure of information, etc. This line of communication ethics, then, could be said to contrast with the "acts and omissions doctrine", which is upheld by some medical ethicists and some religions: it asserts there is a significant moral distinction between acts and deliberate non-actions which lead to the same outcome. In communication ethics, acting and omitting to act would, at first sight, conform to consequentialist theories which hold that a deliberate action is no different from a deliberate decision not to act.

The case of communication is different, however: a silence, for example, is not the mere omission of a message if it is attributed meaning by the other participant(s), and therefore also a communicative act (cf. [15]: "One cannot not communicate").

Arguably, there is no strictly equivalent act (message) that would produce exactly the same outcome in a given situation.

With these last remarks, it has become clear where the limits of the transactional model of communication lie. In the transactional model, the sender (encoder) produces a message or omits to do so and, by doing so, makes ethical decisions about form and content of the message. The receiver (decoder) has no part in the production of the message, but only the responsibility to decode it correctly. As soon as we have to admit that the receiver is actively involved in producing the meaning of a message, the information theoretical model of communication starts to fail and, at the same time, a communication ethics that has to rely on the sender of the message as the lonely decision-maker.

3.2.3 Dialogic and Interpersonal Communication Ethics

Interpersonal communication ethics [16] is different from the former kinds of communication ethics, because it is actually concerned with the relationship between persons: "Interpersonal communication finds its identity in the ethical mandate to protect and promote the good of the relationship" (p. 119). In the case of the authors, this mandate has its roots in a Christian philosophy and is not further discussed. They start their reflections with three assumptions:

1. Interpersonal communication is defined by the primacy it gives to relationship care, so that by far not all dyadic exchanges can be called interpersonal communication, but is not limited to private discourse.
2. Interpersonal communication cares for the relationship in order to "bond responsibility between persons", not to advance goals and plans of individual participants.
3. Interpersonal communication is understood as "relational nurture", with the assumption that relationships need to grow and change. Information, context, or even the speakers are secondary to the relationship as the privileged good.

An important ingredient to relationship care is distance: "relationships need space for growth and change" ([16], p. 123). Interpersonal communication therefore must not be mistaken for an increase in closeness. The assumption of interpersonal communication ethics is that despite the closeness of a relationship, the space for distance and distinctiveness of persons is needed. For example, the distinctive roles of persons in a relationship, e.g. mother and daughter, need to be maintained and cared for as such and should not be given up in favour of a closer, but indistinct relationship, e.g. of friends. Finding and maintaining distance is an ethical responsibility of the participants in a relationship. We find here again, in different words, two issues for an interaction ethics that have already been mentioned: the other's autonomy as a person and her social identity, which both have to be concerns of a participant in social interaction.

The ethical dimension of interpersonal communication moves the focus from relationship to responsibility for the relationship. The notion of responsibility engages a communicator with an ethical charge to attend to practices that bind a given relationship together. This view of interpersonal communication does not put style, liking, or feeling good into the foreground, but instead the responsible engagement. The central question is how to be "good as" the identity which one has in the given relationship, e.g. what it means to be good as a teacher, a mother, a friend, etc.

Arnett's [16] focus on relationship, excluding all task-, action-, or information-oriented communication as objects of study, narrows down extremely the domain of applicability of such an interpersonal ethics. And it begs the question on how to decide which conversations (if any) are eligible and which not.

The assumption that there is a clear boundary between relationship-focused conversations and others that are not is, from the point of view of discourse analysts of any flavour, not applicable to naturally occurring conversation, because every conversation entails both pragmatic and social goals. The other way around, the fact of having a conversation with someone already implies that there is some relationship between the participants. The relationship may not qualify for the high standards of intensity and goals that Arnett et al. have in mind, but we find no hint there where the line would have to be drawn between ethically relevant conversations in their sense and all the rest.

3.2.4 An Ethics of Social Interaction?

Although one has to conclude that Arnett et al. [16] try to define a counterfactual ideal type of conversation, the approach makes a few relevant points in that it:

(a) Acknowledges the conversation as a mutual achievement
(b) Underlines the social and relational character of conversations

It is possible to translate the principles of interpersonal communication ethics into an interaction ethics by giving it both a wider scope and a more modest goal:

1. Care for the relationship is an element of each social interaction, whatever the relationship is.
2. The goal of relational work in social interaction is to foster social bonding (but, curiously, not necessarily between the participants).
3. The ethical mandate for the participant in social interaction is to pursue this goal, while orienting to the other as a person.

Of these, the notion of "social bonding" used here may need some explanation. In our understanding, the need to bond is a fundamental need of social beings such as humans. Humans are highly sensitive and react emotionally to threats to their social bonds (the typical social emotions here being shame and embarrassment). In an evolutionary perspective, the basic "need to belong" makes sense, because solitary human beings were unlikely to survive and reproduce, so that natural selection favoured individuals who are inclined towards sociality and group living. This fundamental need to belong (to bond socially, [17–19]) makes people acutely sensitive to the degree to which they are being accepted by other people. People appear to be particularly attuned to instances of real or potential relational devaluation, i.e. indications that others do not regard their relationship with the individual to be as important, close, or valuable as the individual desires. Leary [20] uses here the notion of a "sociometer", a psychological monitor that scans the social environment for cues indicating social exclusion and which appears to operate largely on a preattentive level, allowing people to devote their conscious attention to other things, but alerting them and claiming attention to interpersonal cues especially of a negative kind (disinterest, disapproval, or rejection). Negative emotions seem to be much more frequent than positive ones in social interaction. While negative emotions appear to be aroused quite easily, positive social emotions require very strong indications of love, respect, or acceptance. This asymmetry was found repeatedly in experimental studies and appears to be neurophysiologically founded.

In the history of sociology, two goals of a person in society turn up again and again, with different focus and names. We will call them here "autonomy" and "bonding". Autonomy encompasses the desire for individuality, independence, freedom, self-determination, etc., while bonding stands here for the needs to belong, to be member of a group, and to be accepted and appreciated by others.

The balance between the two may vary individually and culturally. Totalitarian collectivities, e.g. armies, suppress the need for autonomy almost completely for the sake of membership, but have to compensate for the loss by fostering indirect self-esteem as member, e.g. through uniforms and rituals. Overall, however, humans are assumed to pursue both desires simultaneously and to seek a suitable balance.

Putting the need to belong beside the need for autonomy raises the question why respect for sociality should not be considered a prima facie duty with as much justification as autonomy. This approach, of course, brings to mind the ethics of care [21] which implies that there is moral significance in the fundamental elements of relationships and dependencies in human life. Normatively, care ethics seeks to maintain relationships by contextualising and promoting the well-being of care-givers and care-receivers in a network of social relations. The care ethics perspective on nurturing/caring and dependency relationships however would be too narrow to account for what we have in mind here as an interaction ethics, and it stresses the relationship aspect at the cost of the autonomy aspect.

Interaction ethics rather has to respect both prima facie duties equally, taking into account their potential contradiction. The duty of balancing both, then, could be considered its distinctive feature.

The outline of an interaction ethics given above makes it the probably most mundane and everyday branch of ethics. One could rightly object to it that it does no more than describe what people normally do anyway.

The edges between social and moral norms are indeed blurred. Dubreuil and Grégoire [22] discuss some previous approaches to distinguish social and moral norms and have to conclude that there are many instances where it is hard to give an unambiguous answer. One type of motivation—social or moral—could be dominant under certain conditions and the other entirely irrelevant, but they consider this to be an empirical question that could only be solved by studying norm compliance behaviour. One such empirical study [23] seems to bear them out, because its results show pretty much a continuum of moral/social judgments by subjects. The distinction is even harder for interaction ethics because, although each interaction involves numerous choices and decisions to be made by participants, most of them having consequences for their autonomy and membership needs, these decisions are highly routinised and fast and people are hardly aware of them.

However, if we take awareness as a necessary condition for ethical decisions, thus excluding routine actions, we would introduce a factor that could only be assessed by introspection. It would then be impossible, for the observer, to qualify any decision (rather, the ensuing action which is what can be observed) as ethical. If a physician, for example, executes her duty in line with medical ethics—does she do it routinely or ethically? If a self-driving car does not run over a person—does an ethical system prevent it from doing so, or a low-level process that actuates the brakes before obstacles of whatever kind? For all practical purposes, and in particular for the person who is run over, the reply to both questions would be: "Who cares?"

Social norms are not necessarily ethical, but usually contradictions appear only at the edges of the group having that norm and to external observers. I will leave out this debate at this point, because most contributions in this book tacitly assume that robots—and their ethical systems—are built and used within one and the same civilisation, where most accepted social norms do not go against shared moral principles.

3.3 Ethical Aspects of Social Interaction

In this section, we will illustrate some aspects of conversation that are of relevance to an interaction ethics. Given their ubiquity, it is hard to isolate kinds or elements of interaction where ethical considerations come into play and quite impossible to find examples where they would not. The purpose of this section, then, is to illustrate the different facets of interaction and their particular problems. We will distinguish, for the sake of presentation, three facets: the ""mechanical", the semantic, and the relational, bearing in mind that in actually occurring conversation, they can hardly be separated.

3.3.1 Mechanics: Turn-Taking and Topic Change

As a discipline, Conversational Analysis (CA) lies at the crossroads of sociology with linguistics, social psychology, and anthropology. Its sociological roots lie in Goffman's explorations of the interaction order [50] and Garfinkel's programme of ethnomethodology [24]. One of the key sociological issues that CA addresses is intersubjectivity: how do we share a common understanding of the world and of one another's actions in the world? CA's contribution to this long-standing issue in sociology is to show that it is possible to gain analytic access to the situated achievement of intersubjectivity by focusing on the sequential organisation of talk. CA has had a major influence on linguistics, by making naturally occurring talk a worthy object of study, methodologically through recordings and transcriptions, and conceptually by correcting the then predominant view that actual talk was no more than the deficient application of logical and linguistic laws of language [25, 26].

Conversation analysis is characterised by the views that how talk is produced and how the meanings of talk are determined are the practical, social, and interactional accomplishments of members of a culture. Talk, or "talk-in-interaction" as it is often called in CA, is not simply the product of speakers and hearers who exchange information or send messages to each other. Participants in conversation are seen as mutually orienting to and collaborating in order to achieve orderly and meaningful communication. One aim of CA is to reveal the organised reasoning procedures which inform the production of naturally occurring talk. Sacks et al. [27], the founders of CA, even talked about the "machinery" of conversation to underline the instrumental nature of these procedures, methods, and resources which are available to participants by virtue of their membership in a natural language community [28].

Our former definition of social interaction owes to CA its focus on the interaction as a joint, local, and spontaneous achievement of participants. In this section, we will focus on two phenomena that have been studied in CA to see whether they can involve dimensions of interaction ethics in the sense outlined in the previous section: interruption and topic change.

Interruption

It is the principle of turn-taking that makes a conversation possible. Participants have to achieve that "(1) at least, and no more than, one party speaks at a time in a single conversation, and (2) speaker change recurs". [27]

Interruption refers to a transgressive act in turn-taking where someone starts to speak in the midst of another's turn and prevents the other from finishing it. In more technical CA terms, it means to start a turn at a point which is not a legitimate transition-relevant one. We have to leave out here the background about such points are and how the concept has been explored in CA and only state that CA discovered early on that one participant projects and the other perceives appropriate places for speaker change so that turn-taking can take place in an orderly and consensual way.

One speaker starting a turn while the other speaker is still in the middle of her turn is not per se an offence or a breach of etiquette. There are also "interjacent overlaps" [29] to be taken into account, where speakers start their turns prematurely, but where the overlap is still in line with orderly and cooperative turn-taking practices. This is the case, for example, when the recipient recognises the point of an ongoing turn half-way through it and expresses exactly that by completing it for the speaker or in collaboration with her. Likewise, warnings or hints to the situation may be uttered at any time, even in the midst of another's turn, without them being addressed as interruption by the recipient because of their immediate relevance in the situation. CA then can only treat those cases as interruptions where the recipient orients towards this transgression as such: only when the interrupted participant "does being interrupted" [30], the analyst can assume a transgression. The clearest sign is, of course, when the interrupted party addresses the disturbance explicitly, e.g. "I haven't finished" or "Please don't speak when I'm speaking", where these utterances necessarily are themselves started during the ongoing turn of the other. Quite often, however, addressing the interruption explicitly is only the culmination of a series of "floor usurpations" by a participant which the speaker may accept or try to ignore (e.g. by trying to talk through the interruption).

By its own rules, then, CA cannot possibly reveal cases of systematic and aggressive denial of turn completion when it is not contested by the recipient at least implicitly. Cases where one appears to have the "right to interrupt" by virtue of his or her status and the other acknowledges this right will go unnoticed although they might be more relevant from an ethical point of view.

Topic Change

The commonsense understanding is that the topic is what the conversation is about. Conversation analysts turned their attention more to the structural aspects of topic change [31, 32], a perspective that highlighted topic changes as "a solution to the problem of producing continuous talk" [31, p. 265]. The motivation for this perspective lies in the difficulty to determine, from the conversationalists' standpoint, what the content of a sequence is:

> Disparate topics can occur coherently within the framework of a single, expanded sequence and achieve coherence by being framed by it. An utterance apparently coherent topically with preceding talk can appear incoherent nonetheless if it is structurally anomalous within the sequence it is part of. [33]

The conversation that is the basis for Schegloff's statement shows an intricate intertwining of things that are "talked about" which, nevertheless, are made coherent by the participants. His paper reminds us that the topic of a conversation is whatever the participants jointly construe as the topic, including and excluding things in the process. Although we will stick with the simpler notion of topic as the content of the human–robot dialogue in the following analysis (on the side of the robot, it is technically determined by the currently active dialogue script), this observation still has some influence: a topic can be brought up by one participant (human or robot), but it becomes "the" topic of conversation only by agreement.

The intention of the field study [34] where these data come from was to contribute to filling some gaps in our knowledge about the users of robot companions. We collected audio-visual data on participants' involvement with the robot in their daily lives throughout a period of 10 days. This duration allowed enough time for the novelty to wear off and for routine to build.

The project used a portable set-up with a robotic interface. The robot was always on and so could interact at any point in time, thus constituting a continual social presence in the participant's home. As opposed to a computer interface which has to be turned on to be active and remains a passive responder to user action, the embodied interface is able to actively initiate interactions.

The scenario chosen for the field study was that of a robot companion in the role of a health and fitness coach, with a view to possible practical uses of assistive robots, for example, in rehabilitation. The application was built as consistently as possible around this role so that, e.g., questions and initiatives of the robot were motivated by its concerns and thus understandable for the participant.

For user input, we used yes/no buttons in the first iteration of the study and, for the second and third iterations, the built-in RFID reader of the robotic interface (the Nabaztag, a rabbit-like stationary mini-robot). Participants received a set of flash-cards each equipped with an RFID tag. The cards bore words, numbers, and symbols suitable for the different topics and questions that could occur in the system's dialogues: multi-purpose smileys (looking friendly, neutral, unfriendly) for yes/don't know/no, good/neutral/bad, etc.; numbers for rating (1 to 5), numbers for minutes of activity (10 to 60), and some additional commands and possible topics (message, weather, system, repeat, add-to-log, and, in iteration 3, an "aargh!"/reset/stop card).

Collection of video data was only done when the user consented to it by pressing a button. This button did not start the recording per se (which was instead triggered by the sensors) but did affect whether the video data were permanently stored on the local hard disk. In this way it was possible to record also the beginnings of interaction before the video button was pressed.

The following extracts therefore contain (in the left column) data obtained from the video recordings, the other three columns were obtained from the logfiles. Only the input from the sensors (RFID cards, key hook, and motion sensor) were available to the robot (not the user's talk), and only its speech output was available to the user. In the first extract, the robot reacts to the attempt of the user to introduce a topic ("addtolog", which is a dialogue in which the user can record exercise time to the diary of physical activities).

Extract 1: Rec. no 144, it2_p4May8_0958.mov

Row	User actions (not available to Nabaztag)	User input (available to Nabaztag)	Nabaztag output (available to user)	Nabaztag internal state (not available to user)
1				PIR sensor signal, GOODMORNING dialogue
2			Good morning, how are you?	
3		Card "addtolog"	(blue light)	
4			Before we start talking about that, can I just give you the weather and let you know how much exercise you've got planned today first? Afterwards, please remind me that you wanted to add something to your log	
5	Okay		...	

The robot does not ignore or refuse the user-proposed topic "bald on record" [35] but suggests to postpone it by offering a different sequel of topics, to which the user agrees.

Extract 2: Rec. no 144, it2_p4May8_0958.mov

Row	User actions (not available to Nabaztag)	User input (available to Nabaztag)	Nabaztag output (available to user)	Nabaztag internal state (not available to user)
16			The weather today is rain at first, soon brightening up	
17	Okay, yes, it is brightening up. And now add to log. You said t-			REMINDINGWEIGHING dialogue
18			Some people find it useful to weigh themselves daily as part of their exercise routine	

(continued)

19	No, I will never do that		I hope you don't mind me asking but will you be weighing yourself today?	
20	No, I won't ever be doing that ... But I want to ... (shows card)			

In row 17, P4 shows again that she has gone along with the topic introduced by the rabbit, by repeating part of the last utterance and "okay". Both are ways to close a topic [36, 37], stronger when used in combination. The user immediately opens her topic again, announcing the change with "And now ... " and further legitimating the topic change by referring back to the robot's earlier request "Remind me ... ". Notice that at this point, she does not use the card to launch the addtolog sequence. Possibly she forgets that the robot cannot understand what she says, or else she expects that the robot remembers her topic and comes back to it by itself. Instead, the robot goes on with its own sequence of topics as scripted in the "good morning" dialogue. This denial of her request frustrates the user, her voice becomes louder, and she appears agitated (not represented in this transcript). Her repeated denials to the robot's suggestion turn out to be not just an answer to the specific question, but a refusal of the topic as a whole and in general, and beyond that to a rejection of any topic introduced by the robot at this point which is not her own, as is shown by her renewed effort to change to it. It can be assumed that this topic change by the robot is especially frustrating because the topic of weighing oneself was not mentioned by the robot in its proposed sequence of topics to insert before taking up the addtolog activity. The user seems to have taken the robot's proposal as a promise with which it has committed itself to a sequence of topics including her own. With her "okay" in line 5 (Extract 1), she has accepted the "deal" and now feels betrayed. The robot with its simple dialogue scripts has "succeeded" in offending the human and in being rude (see Sect. 3.3.3 on a further source of rudeness in this dialogue).

3.3.2 Meaning: The Cooperative Principle

This section is concerned with *what* is said in conversation. The contents have long been the realm of logic, on the assumption that the core of human utterances are logical propositions, embedded in acts with which the speaker expresses what he intends to achieve with it [38].

Grice's aim is to see talk as a special case of rational behaviour, and the purpose of his seminal paper *Logic and Conversation* [39] is to show that conversation is guided by logic which becomes visible once one takes into proper account the nature and importance of the conditions which govern conversations. Grice's cooperative principle together with the maxims that it entails (such as "be relevant") is often quoted as guidelines or laws of conversation, acquiring a taste of moral commandment. In this section, we will discuss whether they have been conceived as an "ethics of conversation".

The cooperative principle is introduced by Grice as follows:

> The following may provide a first approximation to a general principle. Our talk exchanges do not normally consist of a succession of disconnected remarks, and would not be rational if they did. They are characteristically, to some degree at least, cooperative efforts; and each participant recognizes in them, to some extent, a common purpose or set of purposes, or at least a mutually accepted direction. This purpose or direction may be fixed from the start (e.g., by an initial proposal of a question for discussion), or it may evolve during the exchange; it may be fairly definite, or it may be so indefinite as to leave very considerable latitude to the participants (as in a casual conversation). But at each stage, SOME possible conversational moves would be excluded as conversationally unsuitable. We might then formulate a rough general principle which participants will be expected (ceteris paribus) to observe, namely: Make your conversational contribution such as is required, at the stage at which it occurs, by the accepted purpose or direction of the talk exchange in which you are engaged. One might label this the COOPERATIVE PRINCIPLE. ([39], p. 45)

From the cooperative principle, Grice derives a number of maxims attributed to the categories of quantity, quality, relation, and manner:

Category	Maxims
Quantity	Make your contribution as informative as is required Do not make your contribution more informative than is required
Quality	Do not say what you believe to be false Do not say that for which you lack adequate evidence
Relation	Be relevant
Manner	Be perspicuous, i.e. Avoid obscurity of expression and ambiguity Be brief and orderly

In particular, the maxims relating to quality all have the flavour of known ethical principles, which may have tempted scholars to view all the maxims as a kind of ten commandments for conversations. For them to be ethical principles, we would expect them to be associated with a value (to follow them is "good") and to be based on intentional decisions of speakers (to be "good" or not). Indeed, some researchers have taken the CP (cooperative principle) and the maxims as rules rather than principles. The difference is that rules can only be obeyed or broken, whereas principles can be upheld to varying degrees. The interpretation as rules leads to the assumption that, following them, one can avoid breaks and mistakes in communication: "Grice's principle assumes that people cooperate in the process of communication in order to reduce misunderstandings". [40] This interpretation made the CP attractive for the design of dialogue systems where mistakes should be avoided wherever possible: the capabilities of repair of such systems are limited if not non-existent (while human–human conversations contain numerous miscommunications which are routinely repaired "on the fly"):

> We conclude that the CP and the maxims, as a necessary side effect of improving understanding and enhancing communication, serve the purpose of preventing the need for clarification and repair metacommunication. [41]

Did Grice actually have such an "ethics of communication" in mind when he formulated the maxims? [42] tries to answer this question by first embedding the paper *Logic and Conversation* (L&C) in the context of Grice's other work. In his philosophy on meaning and language use, Grice is close to the Ordinary Language Philosophers (Austin, Searle) and in contrast with Ideal Language Philosophy (Frege, Russell). An important aim of Grice is therefore a definition of sentence meaning not via truth conditions, but in terms of speaker intention. Put simply, one could say: while Ideal Language Philosophy attempts to adapt language to classical logic and finds everyday language deficient in this respect, Ordinary Language Philosophy wants to discover and describe the logic that already underlies everyday language. In L&C, Grice rises to the challenge of demonstrating the existence of a logic in the operation of conversation, in particular, in the frequent forms of indirectness and non-explicitness that he collects under the term "implicatures". The aim is to show that structures which had evaded the grasp of formal logic could be accounted for in a systematic way and to "see talking as a special case or variety of rational behaviour" (L&C, p. 47)

What is more, not only the philosopher perceives rationality in speakers' and hearers' behaviour, but they themselves have to assume that the other participant behaves rationally in order to understand conversational implicatures. Only on the basis of this assumption does it make sense for the hearer to seek out the speaker's meaning: if we only could expect any nonsense sequence of words in a conversation, we would not even bother to make an effort to understand them.

The concept of rationality is central to Grice's work, as [42] shows. Grandy [43] specifically attributes to him a Kantian notion of rationality. Following this line of interpretation of Grice's work, the CP and the maxims actually acquire a status of moral commandments, as he himself wrote much later [44]:

> Somewhat like moral commandments, these maxims are prevented from being just a disconnected heap of conversational obligations by their dependence on a single supreme Conversational Principle, that of cooperativeness. ([44], p. 370)

Grice also touches on the question what the basis is for the assumption that talkers will in general proceed in conversation in the manner that these principles prescribe: "A dull but, no doubt at a certain level, adequate answer is that it is just a well-recognized empirical fact that people DO behave in these ways; they have learned to do so in childhood and not lost the habit of doing so" ([44], pp. 47–48). But beyond mere habit and factual validity, Grice thinks that, at the basis, the maxims are also rules that we consider reasonable to follow:

> So I would like to be able to show that observance of the CP and maxims is reasonable (rational) along the following lines: that any one who cares about the goals that are central to conversation/communication ... must be expected to have an interest ... in participating in talk exchanges that will be profitable only on the assumption that they are conducted in general accordance with the CP and the maxims. [44]

Relevance

A further example from a human–computer dialogue is presented here as a case of conversational irrelevance which ends in the user's frustration. It is taken from the DARPA Communicator Corpus and quoted from [45]. The system is an early flight ticket booking system:

Extract 3: DARPA Comm. corpus 60941_XX_XX_04_20010608_062510.xml

1	Sys: please remember to speak after the tone. if you get confused at any point you can say start over to cancel your current itinerary. are you calling from a cellular phone?
2	Usr: no
3	Sys: what are your travel plans?
4	Usr: i'd-like-t'go-from duluth:: (0.3) minnesota to chicago-illinois
5	Sys: what month and day do you wish to depart toulouse
6	Usr: november-seco (0.4) twenty o
7	Sys: what time do you want to leave toulouse
8	Usr: (input not recognised)
9	Sys: what time do you want to leave toulouse
10	Usr: in-the-late-morning
11	Sys: if you have a preferred airline please state it now or say no preference
12	Usr: american airlines
13	Sys: let's see then ... i have you going from toulouse minnesota to chicago illinois on november second is that correct
14	Usr: no
15	Sys: oops ... let's try again i have you departing from toulouse minnesota. did i get that right?
16	Usr: yes
17	Sys: i have you arriving in chicago illinois. have i got that right?
18	Usr: yes
19	Sys: you will depart on november second is that correct?
20	Usr: no
21	Sys: what month and day do you wist to depart toulouse
22	Usr: november(0.2)twentyninth
23	Sys: let's see then ... i have you going from toulouse minnesota to chicago illinois on november twenty ninth is that correct?
24	Usr: yes
25	Sys: please let me check availability and see what i can find
26	Sys: no records satisfy your request. would you like to depart on a different date?
27	Usr: no!
	(call continues)

The emphasis in the user's "no" in line 27 is not reproduced here, but it makes clear that she is frustrated. The error in the ASR (automatic speech recognition) occurs when the system processes the user's input in line 4: "Toulouse" instead of "Duluth". Throughout the dialogue, the user does not notice the system's error. Actually, the call continues after this extract with the same error repeated and still unnoticed and more indignation on the user's side. Considerable effort and patience on the user's side go into giving and correcting input without it being of any use because the first input given is wrong and the rest of the dialogue is built on wrong assumptions. Human participants in conversation expect that each piece of common ground achieved is secured before moving on to the next. Hence, when the system goes on to the next question in line 5, this is taken by the user as an implicit acknowledgment that the previous information has been taken in by the system and accepted as plausible. This assumption is further confirmed by each following question–answer pair, so that, in the end, one is not surprised that the user does still not detect the original error made.

The error in itself is quite harmless, misunderstandings happen all the time in human conversations, and repair activities are commonplace. What is noticeable here is that the system fails to check immediately the plausibility of the user's input: a comparison with the list of airports would show that, in fact, there is no "Toulouse, Minnesota". This failure makes the following sequence of questions and answers irrelevant, and the user reacts emotionally and offended to this breach of the maxim, "Be relevant!".

At least this maxim, then, can be said to have moral underpinnings. The hearer constructs meaning on the assumption that the utterance is relevant to him or her, i.e. that the speaker relates to the hearer. This disposition to find sequential relevance [46] leads participants to try to construct a coherent meaning even out of sequences like

```
A: Is there a gas station around here?
B: It's Sunday.
```

The reader likewise will readily and easily imagine a context where this exchange makes perfect sense. Where it becomes impossible to understand an utterance on these grounds, it is taken not just as the result of miscommunication, but as an offence to the hearer's needs.

Description

Before turning to the discussion of these relational wants, there remains a phenomenon regarding contents of talk to mention which is not captured by Grice's maxims. The reason may be that he was not concerned with lexical semantics, but with sentence meaning, and took shared and uncontested word sense tacitly as a precondition of his principles.

Even when speakers describe commonplace events they have a wide range of alternative possibilities from which to choose, and they choose their words with a purpose. The ethical implications of this become obvious in the following example (from [47], reproduced in [48]). In this transcript of a rape trial, the counsel for the defence (C) is cross-examining the prosecution's main witness (W) who is the victim of the alleged rape. The two extracts are given in a simplified transcription:

Extract 4 (Drew, 1992: 489); numbers in brackets mean pauses in seconds

```
C: [referring to a club where the defendant and the victim
       met] it's where uh .. uh girls and fellas meet isn't it?
(0.9)
W: People go there.
[conversation continues ...]
C: An during the evening (0.6) uh: didn't mistuh [name] come
      over tuh sit with you
(0.8)
W: Sat at our table.
```

In the first extract, the counsel describes the club as a place where young people from both sexes go to make contact with the opposite sex, thereby inviting the inference that the victim went there with the intention to make contact with men with a view to sexual relations and that the men present there, in particular the defendant, could legitimately expect this to be her intention. The witness makes it clear that she understands these inferences and, after a pause, comes up with her own description which is as neutral as possible: "people" leaves out the gender of the club's visitors, and "go there" does not imply any intentions. Similarly, in the second extract C alleges to an existing relationship between the victim and the defendant by using the description "sit with you". Again, the victim is able to draw a totally different picture in two respects: first, she presents herself as a member of a group ("our") and, second, "sat at" does not imply a personal relationship of the man in question with anyone of this group.

The pauses (which are relatively long for conversations) are telling: dispreferred responses are generally accompanied by hesitation and delay [46]. Many types of turns have "preferred" next turns, e.g. questions, invitations, and greetings. The preferred sequel to a yes/no question is obviously "yes" or "no"—with the interesting tendency of speakers to formulate questions where a "yes" can be rather expected than a "no". The same is true—as is the case here—for descriptive statements that beg for confirmation. The problem for W in these extracts is that the dispreferred answer "no" would deny only the statement but would accept its connotations. The victim cannot deny the statement—she did go to the club, and the man came to the table—but wants to contest the description. The long pause is indicative of this even more dispreferred answer which changes the whole frame of reference.

This example makes sufficiently clear that factual descriptions can have ethical implications, but courtroom trials are only the most obvious cases. The (unequally distributed) potential to "name the world" and its moral consequences have been the object of study of Critical Discourse Analysis which, however, is more often concerned with mass media texts than with naturally occurring conversations [49].

3.3.3 Relationship: Politeness

The standard work on politeness is [35] (Brown and Levinson, subsequently cited as B&L). They take up Goffman's [55] concept of "face" in construing politeness as the avoidance, reduction, or compensation, etc., of a "face threat". Goffman defined face as the positive social value a person effectively claims for himself by the line others assume he has taken during a particular contact: "... the person tends to conduct himself during an encounter so as to maintain both his own face and the face of other participants".

This makes interaction dependent on a reciprocity of perspectives by which each respects the self-presentation of the other in expectation of being accorded the same respect [51]. B&L's notion of face also draws on the folk psychological use as in "saving" or "losing" face. Face is emotionally invested and can be lost, maintained, or enhanced, so that it must be constantly attended to in interaction. Normally everyone's face depends on everyone else's being maintained, and

> since people can be expected to defend their faces if threatened, and in defending their own to threaten others' faces, it is in general in very participant's best interest to maintain each others' face. [35]

The assumption on which they start their book is "that all competent adult members of a society have (and know each other to have) 'face', the public self-image that every member wants to claim for himself" [35]. They construct face as having two related aspects:

(a) Negative face: "the basic claim to territories, personal preserves, rights to non-distraction—i.e. to freedom of action and freedom from imposition"
(b) Positive face: "the positive consistent self-image or 'personality' claimed by interactants", which includes the desire that this self-image be appreciated and approved of

In order to derive (linguistic) actions from these concepts, B&L re-define them as individual wants, so that negative face is the want of a member of society that his actions be unimpeded by others, and positive face the want that his wants be desirable to at least some others. The definition of positive face appears somewhat strained, and the authors make an extra effort to explain that these may also be past goals that have already been achieved, so that the person's want is to have her achievements appreciated by others. For present purposes, the simpler way, namely, to say that positive face is the want to have one's self-image appreciated by others is sufficient.

The relationship between these notions of face and what we have called the basic needs of individuals—autonomy and sociality—is obvious, although the notion of positive face is defined here in more individualistic terms than we have done with the notion of "need to belong". That is, maybe, why Culpepper [52] specifies different aspects of positive face and adds "sociality rights" to his list of targets affected by impoliteness:

- (face) quality face: we have the fundamental desire for people to evaluate us positively in terms of our personal qualities (e.g. appearance, abilities, dignity).
- (face) relational face: we have a fundamental desire for people to acknowledge and uphold our relationships with significant others (e.g. closeness/distance, love, trust).
- (face) social identity face: we have a fundamental desire for people to acknowledge and uphold our social identities or roles (e.g. as team leader, teacher).
- (sociality right) equity: people have a fundamental expectation that they are entitled to personal consideration form others and to be treated fairly.
- (sociality right) association: people have a fundamental expectation that they are entitled to an association with others that is in keeping with the type of relationship that they have with them (involvement, shared concerns/feelings, respect).

The addition of a distinct sociality aspects reminds of identity theory [53], where three layers of identity are commonly defined: person, role, and social identities. A social identity based on membership in a group or category gives one self-meanings that are shared with others. By contrast, a role identity is tied to other members of the role set, especially its complementary counterrole (e.g. teacher–student or father–child). The person identity refers to the individual as a bio-social being: the person distinguishes himself or herself as a unique, identifiable individual. These identities are present and need to be verified simultaneously in every interaction. If in this chapter we have chosen to stick with the two dimensions of autonomy and sociality, it is not because we negate this threefold distinction but simply out of the desire not to complicate things further: it is possible, for the present discussion, to understand a role as membership in a group, albeit an idealised and virtual one.

Interestingly, all of Culpepper's categories can be traced back to a concept of positive face if taken in a broad sense. Negative face or any concept similar to it is absent from this study on impoliteness which took and analysed as data narratives of events reported as impolite or rude by subjects from different cultures. It remains open whether the data in fact do not contain any reports of transgressions on negative face or whether the categorisation of the data led to this result. It may also be speculated that such attacks would not be experienced as "impolite" by the informants in any of the meanings that they give this term, but as something else which we ignore.

In B&L, focusing on politeness rather than impoliteness, there is no mention of actual face attacks or face aggravations, but only of face-threatening acts (FTA). An FTA is any act that by its nature runs contrary to the face wants of the addressee and/or of the speaker. According to this definition, one arrives at the following two by two matrix of types of FTA:

	Threat to positive face	Threat to negative face
Of speaker	Apologies Acceptance of a compliment Breakdown of physical control over body Self-humiliation Confessions, admissions of guilt	Thanks and acceptance of thanks or apologies, excuses, acceptance of offers, unwilling promises
Of hearer	Disapproval, criticism, contempt Ridicule, insults, accusations Contradictions, disagreements, challenges Irreverence Non-cooperation, non-attention	Orders and requests Advice, suggestions Remindings Threats, warnings

A further assumption—beside face—their work is based on has to be that of a rational agent who, in the context of mutual vulnerability of face, will seek to avoid these FTAs or will employ strategies to minimise the threat. B&L conceive of this as of a process of weighing several wants:

The want to communicate the content
The want to be efficient or urgent
The want to maintain H's face to any degree
Plus (but not mentioned at this point in the book), the want to maintain S's face. S then has a set of possible strategies (Fig. 3.2).

Roughly, "on record" means that the FTA is expressed unambiguously (e.g. "I request you to ... "), the intention is made clear whereby the speaker commits herself to the FTA, while "off record" leaves the intention ambiguous, so that the speaker cannot be held accountable for the FTA (e.g. "Oh, damn, I forgot my purse!" which may, but need not be, intended as a request to the hearer to pay for the speaker). Doing the FTA baldly without redress involves the most direct,

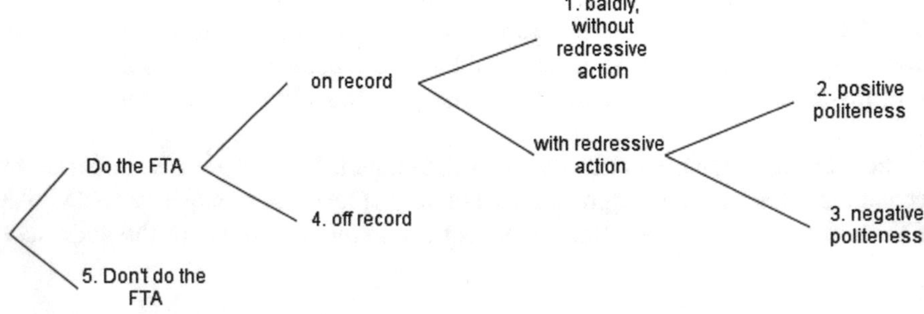

Fig. 3.2 Possible strategies for doing FTAs (B&L, p. 69)

clear, and concise way possible (e.g. "do X!"), while redressive action is any action that attempts to counteract the potential face damage. The redressive action may address positive face (positive politeness) by enhancing the hearer's self-image (status, group membership, etc.), indicating reciprocity, closeness, respect, etc., or it may address negative face (negative politeness) by minimising interference with the hearer's freedom and self-determination, and by indicating a face-saving "way out" for the hearer.

Can Robots Be Rude?

In other words, does what robots do have an influence on the face of the user? In most definitions of rudeness (and of unethical actions in general), the intention of the actor is an important criterion. Culpepper [52] reports that in numerous cases, the subjects feel offended and hurt although they know that the other had not done it (fully) intentionally. It seems that people see intentionality as a scalar attribute, not as an on/off criterion. This is less strange than it appears at first if one considers the explanations that subjects gave about the "partial intentionality", for example: "he didn't do it intentionally, but he could have foreseen that I'd be offended".

Given also that we know that users take (at least sometimes, at least partly) an "intentional stance" [54] towards robots, these two things together lead us to the answer that yes, robots can be rude.

This statement has further to be qualified: rudeness/impoliteness is also, or mainly, in the eyes of the beholder (or the ear of the hearer). It is not (only) a category of action, but (also) of experience. As for everything that happens in interaction, it takes at least two to establish the meaning of any discursive act. The modified statement then reads: robots' actions can be experienced as rude.

What has to be added—and doesn't make things easier in any way—is the situation in which the "rude" act takes place. This ranges from the immediate context of the conversation to the state of a long-term relationship, from external circumstances (e.g. the presence of a bystander, the time of day, etc.) to the mood of the human participant. From the data of the SERA field study ([34], see Sect. 3.1), we have examples of interactions which illustrate the individual responses of the subjects [6]. The suggestion by the robot to the users to weigh themselves was introduced in the robot's script with the purpose of gathering evidence about a scenario that is regularly at issue when socially assistive robots are discussed [1, 7]: what happens when the robot has to do something that is unpleasant to the user, but necessary for his/her health or safety? How should the robot go about it, and how do users react?

Extracts 5 and 6 are from the first iteration of the study described in [16, 34], with a direct question:

Extract 5: Rec. no 5, it1_P1_PSeP26_0948

```
N: Please press the no or yes button to get your answer.
   Have you weighed yourself yet today?
P1: Yes (presses button simultaneously)
N: Okay. Thanks
P1: (smiles)
P1: anything else. I even know I weighed 82.7 kilos or in
    english 12 10 and a half pounds - well stones and pounds.
```

Extract 6: Rec. no. 14, it1_p1_PSep27_0849.mov

```
N: recording on. The weather forecast is mainly dry with
     variable cloud amounts and some sunshine. Have you
     weighed yourself yet today?
P1: Oh, yes! (presses button simultaneously, frustrated
     expression)
N: Okay. Thanks.
P1: and I'm still twelve stone ten and a quarter. (slightly
      irritated voice, turns away)
```

Subjects did have no doubts that the question was to be understood indirectly as a request or reminder. Subject P1 was the only one who complied with it at least initially. Her reaction in extract 6 seems to indicate, however, that the issue we chose as a reminder is an FTA not only to negative face: in our society, the capability of managing one's own weight is considered as part of the self-image, hence of positive face. The robot's question is understood not only as a request here, but as a criticism of her lifestyle or self-discipline, hence a double FTA.

In later iterations of the experiment, the dialogue was modified to be both more polite, employing several politeness strategies at once:

Extract 7: Rec. no 197, it3_p6_FNov6_0834s33.mov

```
N: Some people find it useful to weigh themselves daily as
     part of their exercise routine. I hope you don't mind me
     asking but will you be weighing yourself today?
P6: (shows card) NEUTRAL
```

Extract 8: Rec. no 144, it2_p4May8_0958.mov

```
N: the weather today is rain at first, soon brightening up.
P4: okay, yes, it is brightening up. And now I've to log.
      You said ...
N: some people find it useful to weigh themselves daily as
      part of their exercise routine.
P4 (barges in): no, I won't never do that.
N: I hope you don't mind me asking but will you be weighing
       yourself today?
P4: No, I won't (ever?) be doing that. ... But I want to ...
       (card ADDTOLOG)
```

Extract 9: Rec. no. 160, it3_p2_FOct8_0746s2.mov

```
N: Some people find it useful to weigh themselves daily as
     part of their exercise routine. I hope you don't mind me
     asking but will you be weighing yourself today?
P2: (shows card) FROWN
(P2 half-mouths the weighing question along with the rabbit,
     in a slightly exasperated way.)
```

The politeness strategies used in the robot's script were (B&L, p. 131):

- Stating the request as a general rule: "Some people find it useful ... "
- Deference and apology: "I hope you don't mind me ... "
- Indirectness (off-record), and leaving a way out: "Will you be weighing ... "

P6 (Extract 7) takes this way out by orienting towards the question directly without accepting its suggestive allusion. Extract 8 is actually the same recording as in Extract 2, and we have discussed there the robot's rudeness in violating good manners of floor management, to which it adds now the FTA of the weighing question. The repeated, increasingly emphatic rejection by the subject orients towards this multiple attack on her face: "no ... never ... no ... I won't ever". It is impossible to extract precisely the part that the nature of the FTA (allusion to weight and weight management, request) plays in this insult, but from other records of this subject it is obvious that it has an impact by itself. In Extract 9, also just one of a series of similar interactions, we are led to assume that the FTA is predominantly directed against negative face. What is not certain about this case is, however, whether it is the reminder/request character that is interpreted as face-threatening or whether it is the mere fact that the robot repeats this dialogue every morning with the same words: people tend to react emotionally when they feel that their time is wasted. We can express this in terms of Grice's maxims, namely, that of quantity, or in the terms of an attack on negative face in that being able to dispose of one's time is part of one's wants of self-determination.

These examples were presented here also to illustrate the vast gap between constructed lab interactions with robots and those that actually take place out there "in the wild". Robots that were designed to be polite and friendly are experienced as rude for sometimes unexpected reasons and not least through the active contributions of the human users. Extract 7 is just one out of many similar interactions that subject P6 had with the robot which were all smooth, successful, and calm. He was one of two participants who adapted to the robot's timing, poor capabilities, and conversational routines in such a way that they ended up with a "polite" robot although the technical set-up was identical with that of the "rude" ones.

Autonomy, Sociality, and Interaction Ethics

With these findings in mind, we can return to the example of an ethical eldercare robot [7, 8]:

> For instance, it might be the case that ... harm is present in the action of not notifying an overseer that an eldercare robot's charge is refusing to take his/her medicine. ... it might also be the case that not notifying an overseer exhibits the presence ... of respect for autonomy.

The proposed solution of ethical reasoning leads to a clear result, based on a calculus of preferences: do A, don't do B, because A has priority in a given situation. Interaction ethics, however, tells us that ethical reasoning is not a yes/no decision, but a choice among a wide range of alternatives of how to carry out the action (or non-action) that follows from the original decision. In other words: if a threat to the autonomy of the person can be foreseen, and the person is, in that situation, a competent actor, then redress in social interaction as much of the threat as possible.

One precondition for employing face-saving strategies is that the building-up of the conflict over time is represented in the robot's system. A human caregiver in our case would "see it coming" and make sure that the patient does so, too. She would use a battery of different strategies—utter reminders, requests, arguments, warnings, threats, etc., with growing force and urgency. As a last resort, apologies for the transgression of autonomy rights are in order. The transgression has to be done anyway, but to human beings it makes a lot of difference how it is done.

Anderson and Anderson [8] develops this scenario further, and we are grateful for an example of a situation where the other side of politeness—positive face—comes into play:

> Early in the morning EthEl [the care robot in an assisted living facility] stands in a corner, plugged into the wall socket. Once her batteries fill up, her duty of beneficence ("do good") overrides her duty to maintain herself, so she starts making her way around the room, visiting residents and asking if she can be helpful in some way – get a drink, take a message to another resident, and so on. As she receives tasks to perform, she assigns initial levels of satisfaction and violation to each duty involved in the task. One resident, in distress, asks her to seek a nurse. Ignoring the distress of a resident means violating the duty of nonmaleficence ("prevent harm"). That duty now overrides her duty of beneficence, so she seeks a nurse to inform her that a resident is in need of her services. Once this task is completed, her duty of beneficence takes over again, and she resumes her rounds.

Now imagine that at the moment where the emergency call is received, the robot is performing a service to another resident, as most of the services mentioned here are of social assistance. From a viewpoint of interaction ethics, the robot also has a duty towards any resident with whom it is interacting at the moment. Let's call it a duty of sociality ("be social"), which dictates that the social needs and wants of the person with which the robot is interacting at any one moment have to be cared for. Even with the emergency call taking immediate precedence over this duty of sociality, an interaction-aware ethical system will calculate the harm that will be done by suddenly leaving the current social partner alone, and generate polite behaviour to mitigate the offence (e.g. apologies, explanations, a promise to be back).

3.4 Summary: The Need for a Robot's Social–Ethical System

This chapter has dealt with ethical decision on, so to speak, a microscopic scale. While most ethical reflection, also when it comes to a robot's ethical system, is concerned with decisions on acting or not-acting on the large scale of rare and highly visible actions, our microscopic view breaks down these big deeds to a complex sequence of actions with which any ethical decision has to be carried out. Many of these micro-actions are social, given that they involve interacting with human beings in their execution. These social interactions cannot be reduced to sending a message, which is why communication ethics, as it is commonly understood, is not sufficient: it is limited to the respective duties of speakers and hearers in producing and receiving messages, but cannot do justice to the joint undertaking of leading a conversation.

Any conversation requires, from all participants, constant efforts on three layers:

- The mechanical layer: a conversation has to be started, maintained, controlled, and ended in an orderly way
- The meaning: participants have to cooperate in order to achieve a shared understanding
- The relationship: participants have to respect and confirm their own and the others' face

In conversations among humans, this requires nothing but "normal" behaviour that has been learned from early on. In conversational robots, these duties are mostly assigned to dialogue management, which designers try—not overly successfully so far—to adapt to expected normative behaviour. In this chapter, we have tried to show that interaction management can benefit from ethical reasoning which goes all the way down from "big" decisions to the micro-level of dialogue planning and language understanding.

Based on the simple statements that

- The ethical agent is not alone, but embedded in a social world
- Ethical decisions mostly regard people
- The actions following from ethical decisions have their own ethical dimensions

we arrive at the logical consequence that a robot needs an integrated social–ethical system. The specific features of the social–ethical system are:

(a) Ethical evaluation and decision-making are taken beyond action selection into the detail of behaviour generation.
(b) A historical representation of ethical states and conflicts. Not only the final outcome of ethical reasoning may lead to action, but already the build-up of

an ethical conflict and the changes in the balance of ethical priorities (which are translated, for example, in reminders and warnings)

(c) A user (or general: human) model representing (social) needs and desires against which the potential impact of actions can be predicted.

Instead of arbitrary attempts to mimic normative social behaviour and "good manners" in human–machine dialogue, a social–ethical system can evaluate both the human's and the robot's utterances with respect to their expression of and impact on the human's needs and desires.

The line between socially normative and moral behaviour is thin and vague, even for humans. From the viewpoint of a social–ethical robot, normative behaviour would be moral behaviour that has become routine through growing up and being socialised as a human being. As long as robots don't have the same capacities to learn and develop their social capabilities, they may as well use their (superior) reasoning capacities to compensate this deficiency with ethical reflection and decision-making.

References

1. Feil-Seifer, D., Mataric, M.: Defining socially assistive robotics. In: Proceedings of the IEEE 9th International Conference on Rehabilitation Robotics. Chicago (2005)
2. Fong, T., Nourbakhsh, I., Dautenhahn, K.: A survey of socially interactive robots. Robot. Auton. Syst. **42**(3–4), 143–166 (2003)
3. Rummel, R.J.: Understanding Conflict and War, vol. 2. Sage, Beverly Hills (1976)
4. Searle, J.R.: Minds, brains and programs. Behav. Brain Sci. **3**(3), 417–457 (1980)
5. Nass, C., Brave, S.: Wired for Speech. MIT Press, Cambridge (2005)
6. Payr, S.: Virtual butlers and real people: styles and practices in long-term use of a companion. In: Trappl, R. (ed.) Virtual Butlers: The Making of. Springer, Heidelberg (2013)
7. Anderson, M., Anderson, S.L.: Case-supported principle-based behavior paradigm. In: Trappl, R. (ed.) A Construction Manual for Robots' Ethical Systems: Requirements, Methods, Implementations. Springer, Heidelberg (2015)
8. Anderson, M., Anderson, S.L.: Robot Be Good. Scientific American, October 2010
9. Barnlund, D.C.: A transactional model of communication. Transaction. In: Mortensen, C.D. (ed.) Communication Theory, pp. 47–57. Transaction, New Brunswick (2008)
10. NCA, Legislative Council: NCA Credo for Ethical Communication. Edited by National Communication Association. https://www.natcom.org/Tertiary.aspx?id=2119 (1999)
11. Andersen, K.E.: A conversation about communication ethics with Kenneth E. Andersen. In: Arneson, P. (ed.) Exploring Communication Ethics: Interviews with Influential Scholars in the Field, pp. 131–142. Peter Lang, New York (2007)
12. Wrench, J.S., Punyanunt-Carter, N.: An Introduction to Organizational Communication. http://2012books.lardbucket.org/books/an-introduction-to-organizational-communication/ (2012)
13. McCroskey, J.C., Wrench, J.S., Richmond, V.P.: Principles of Public Speaking. The College Network, Indianapolis (2003)
14. Redding, W.C.: Ethics and the study of organizational communication: when will we wake up? In: Jaksa, J.A., Pritchard, M.S. (eds.) Responsible Communication: Ethical Issues in Business, Industry, and the Professions, pp. 17–40. Hampton, Cresskill (1996)
15. Watzlawick, P., Bavelas, J.B., Jackson, D.D.: Pragmatics of Human Communication. W. W. Norton & Company, New York (1967)

16. Arnett, R.C., Harden Fritz, J.M., Bell, L.M.: Communication Ethics Literacy: Dialogue and Difference. Sage, London (2009)
17. Fearon, D.S.: The bond threat sequence: discourse evidence for the systematic interdependence of shame and social relationships. In: Tiedens, L.Z., Leach, C.W. (eds.) The Social Life of Emotions, pp. 64–86. Cambridge University Press, Cambridge (2004)
18. Scheff, T.: Microsociology. Discourse, Emotion, and Social Structure. University of Chicago Press, Chicago (1990)
19. Kemper, T.D.: Social relations and emotions: a structural approach. In: Kemper, T.D. (ed.) Research Agendas in the Sociology of Emotions, pp. 207–236. State University of New York Press, New York (1990)
20. Leary, M.R.: Affect, cognition, and the social emotions. In: Forgas, J.P. (ed.) Feeling and Thinking, pp. 331–356. Cambridge University Press, Cambridge (2000)
21. Sander-Staudt, M.: Care Ethics. Internet Encyclopedia of Philosophy. http://www.iep.utm.edu/care-eth/ (last visited May 2, 2014)
22. Dubreuil, B., Grégoire, J.-F.: Are moral norms distinct from social norms? A critical assessment of Jon Elster and Cristina Bicchieri. Theory Decis. **75**(1), 137–152 (2013)
23. Huebner, B., Lee, J.L., Hauser, M.D.: The moral-conventional distinction in mature moral competence. J. Cogn. Cult. **10**, 1–26 (2010)
24. Garfinkel, H.: Studies in Ethnomethodology. Polity Press, Cambridge (1967)
25. ten Have, P.: Doing Conversation Analysis. Sage, London (1999)
26. ten Have, P.: Conversation analysis versus other approaches to discourse. Forum Qual. Soc. Res. **7**(2) (2006)
27. Sacks, H., Schegloff, E., Jefferson, G.: A simplest systematic for the organization of turn-taking for conversation. Language **50**(4), 696–735 (1974)
28. Eggins, S., Slade, D.: Analysing Casual Conversation. Equinox, London (1997)
29. Jefferson, G.: Notes on ‘latency’ in overlap onset. Hum. Stud. **9**(2/3), 153–183 (1986)
30. Hutchby, I.: Participants’ orientations to interruptions, rudeness and other impolite acts in talk-in-interaction. J. Politeness Res. **4**, 221–241 (2008)
31. Maynard, D.W.: Placement of topic changes in conversation. Semiotica **30**(3/4), 263–290 (1980)
32. Okamoto, D.G., Smith-Lovin, L.: Changing the subject: gender, status, and the dynamics of topic change. Am. Sociol. Rev. **66**(6), 852–873 (2001)
33. Schegloff, E.A.: On the organization of sequences as a source of “Coherence” in talk-in-interaction. In: Dorval, B. (ed.) Conversational Organization and its Development, pp. 51–77. Ablex, Norwood (1990)
34. Creer, S., Cunningham, S., Hawley, M., Wallis, P.: Describing the interactive domestic robot setup for the SERA project. Appl. Artif. Intell. **25**(6), 445–473 (2011)
35. Brown, P., Levinson, S.C.: Some Universals in Language Usage. Cambridge University Press, Cambridge (1987)
36. Howe, M.: Collaboration on topic change in conversation. In: Ichihashi, K., Linn, M.S. (eds.) Kansas Working Papers in Linguistics, pp. 1–14. University of Kansas (1991)
37. Beach, W.A.: Conversation analysis: “Okay” as a clue for understanding consequentiality. In: Sigman, S.J. (ed.) The Consequentiality of Communication, pp. 121–162. Lawrence Erlbaum Ass, Hillsdale (1995)
38. Searle, J.R.: Speech Acts. An Essay in the Philosophy of Language. Cambridge University Press, Cambridge (1969)
39. Grice, H.P.: Logic and conversation. In: Cole, P., Morgan, J. (eds.) Syntax and Semantics. Academic, New York (1975)
40. Finch, G.: Linguistic Terms and Concepts. Macmillan, London (2000)
41. Bernsen, N.O., Dybkjer, H., Dybkjer, L.: Cooperativity in human–machine and human–human spoken dialogue. Discourse Process. **21**, 213–236 (1996)
42. Davies, B.: Grice’s Cooperative Principle: getting the meaning across. In: Leeds Working Papers in Linguistics and Phonetics, vol. 8, pp. 361–378 (2000)
43. Grandy, R.E.: On Grice on language. J. Philos. **86**, 514–525 (1989)

44. Grice, H.P.: Studies in the Way of Words. Harvard University Press, Cambridge (1989)
45. Wallis, P.: Revisiting the DARPA communicator data using conversation analysis. Interact. Stud. **9**(3), 434–457 (2008)
46. Seedhouse, P.: The Interactional Architecture of the Language Classroom. A Conversation Analysis Perspective. Blackwell, Oxford (2004)
47. Drew, P.: Contested evidence in court-room cross-examination: the case of a trial for rape. In: Drew, P., Heritage, J. (eds.) Talk at Work: Interaction in Institutional Settings, pp. 470–520. Cambridge University Press, Cambridge (1992)
48. Hutchby, I., Wooffitt, R.: Conversation Analysis, 2nd edn. Polity Press, Cambridge (2008)
49. Fairclough, N.: Language and Power, 2nd edn. Longman, Harlow (2001)
50. Goffman, E.: Interaction Rituals: Essays on Face-to-Face Interaction. Pantheon, New York (1967)
51. Malone, M.J.: Worlds of Talk. The Presentation of Self in Everyday Conversation. Polity Press, Cambridge (1997)
52. Culpepper, J.: Impoliteness. Using Language to Cause Offence. Cambridge University Press, Cambridge (2011)
53. Burke, P.J., Stets, J.E.: Identity Theory. Oxford University Press, Oxford (2009)
54. Dennett, D.C.: The Intentional Stance. MIT Press, Cambridge (1987)
55. Goffman, E.: The Presentation of Self in Everyday Life. Doubleday, New York (1959)

Chapter 4
Shall I Show You Some Other Shirts Too? The Psychology and Ethics of Persuasive Robots

Jaap Ham and Andreas Spahn

Abstract Social robots may provide a solution to various societal challenges (e.g. the aging society, unhealthy lifestyles, sustainability). In the current contribution, we argue that crucial in the interactions of social robots with humans is that social robots are always created to some extent to influence the human: Persuasive robots might (very powerfully) persuade human agents to behave in specific ways, by giving information, providing feedback and taking over actions, serving social values (e.g. sustainability) or goals of the user (e.g. therapy adherence), but they might also serve goals of their owners (e.g. selling products). The success of persuasive robots depends on the integration of sound technology, effective persuasive principles and careful attention to ethical considerations. The current chapter brings together psychological and ethical expertise to investigate how persuasive robots can influence human behaviour and thinking in a way that is (1) morally acceptable (focusing on user autonomy, using deontological theories as a starting point for ethical evaluation) and (2) psychologically effective (focusing on effectiveness of persuasive strategies). These insights will be combined in a case study analysing the moral acceptability of persuasive strategies that a persuasive robot might employ while serving as a clothing store clerk.

Keywords Persuasive technology • Social robots • Persuasive robots • Psychology • Persuasive strategies • Ethics • Discourse ethics • Deontology

---

J. Ham (✉)
Department of Human-Technology Interaction, Eindhoven University of Technology, Den Dolech 2, 5600 MB Eindhoven, The Netherlands
e-mail: j.r.c.ham@tue.nl

A. Spahn
Department of Philosophy and Ethics of Technology, Eindhoven University of Technology, Den Dolech 2, 5600 MB Eindhoven, The Netherlands
e-mail: a.spahn@tue.nl

R. Trappl (ed.), *A Construction Manual for Robots' Ethical Systems*, Cognitive Technologies, DOI 10.1007/978-3-319-21548-8_4

4.1 Introduction

Social robots may provide a solution to challenges such as an aging society, take over personnel shortages or mundane and repetitive household chores, provide infotainment or support humans in many other ways. The feasibility of social robots that service or assist people has to some extent already been demonstrated in laboratories, but many challenges still need to be addressed. These challenges include gaining a better understanding of the interaction between humans and robots, an issue that is closely connected to technical challenges related to integrating diverse and complex software components. So, social robots that enter the homes, offices and other areas of the daily lives of humans find applications in many and diverse domains.

Crucial in the interactions of social robots with humans is that social robots are always created to some extent to influence humans. Whether it is actual behaviour change (e.g. help humans buy a new shirt or help humans to take their medication) or a more cognitive effect like attitude change (inform humans about danger) or even changes in cognitive processing (help humans learn better), effective influencing is fundamental to social robots.

Indeed, research suggests that technology in general might be very well suited to influence human behaviour or thinking. Persuasive technology (for an overview, see [1–3]) is a class of technologies that are intentionally designed to change a person's behaviour, attitude or both [1, 2]. Importantly and in contrast to what the label persuasive technology suggests, the definition of persuasive technology remains silent about the nature of the cognitive processes that lead to these changes. That is, persuasive technologies do not solely use persuasion in the sense that people change their opinion (and thereby their behaviour) based on arguments and argumentation. Also, persuasive technology might influence users through cognitive processes that are not directly related to persuasion, like classical conditioning, or, for example, by changing the user's perception of social norms.

In the current chapter, we argue that persuasive technology can be very effective when it takes on the form of an embodied social agent. That is, a social robot might be a very powerful (technological) persuader. We argue so because research known as 'the media equation' described by Reeves and Nass [4] indicates that people respond with similar social behaviour when interacting with computer systems as when interacting with humans. Media equation research indicates that when a computer system praises the user (e.g. by saying 'You look beautiful in that shirt'), people make more positive judgments about that artificial system (e.g. judge it to be more attractive) and more positive judgments about working with that computer system [5]. As media equation research [4] indicates that people respond in a similar fashion to computer agents as to human agents, it seems to be the case that social cues (e.g. embodiment, having a voice, facial expressions like smiling or frowning) will lead to automatic social responses [5]. We argue that when an artificial system activates comparable responses as might a human persuader, this artificial system will, just as human persuaders, be able to cause behaviour changes in a person who is given social feedback by that artificial system.

4.1.1 What Are Persuasive Robots?

Recent research started to investigate persuasive robotics (e.g. [6–8]), aiming at gaining an understanding of how robots that will be able to effectively change human behaviour and attitudes can be designed. This research needs to investigate how robots can be effective persuaders, which persuasive mechanisms and strategies robots might effectively use, the characteristics that a robot needs to have for it to be able to employ those persuasive mechanisms and even potential detrimental responses to robotic persuasion fuelled by reactance [9, 10].

In line with earlier definitions of persuasive technology [1, 2], we [7] defined persuasive robotics as the scientific study of artificial, embodied agents (robots) that are intentionally designed to change a person's behaviour, attitudes and/ or cognitive processes [see also 6]. In general, artificial, embodied agents may be very effective persuaders, and the current research contributes to the scientific literature an analysis of how and when they can influence user behaviour effectively in morally acceptable way.

4.1.2 Ethical Concerns About Social Persuasive Robots

The introduction of persuasive technology and persuasive robots, however, changes the relation between human and technology and introduces a new way of influencing human behaviour, by the explicit attempt to design social robots whose function is behaviour or attitude change in the user. This raises important ethical questions. On the one hand, persuasive technologies go beyond the paradigm of technology as neutral tools. Robots that aim at influencing behaviour are inherently normative as they try to steer the behaviour of the user into a 'desired' direction. But the question arises: Who defines what desired behaviour is? Does the user agree with the aims of the social robot and thus use this technology just as a better means to reach his or her end or is there at times a conflict between the aims of the persuasive technology and the user? A robot that helps a patient to remember to take his medicine is prima facie less morally problematic than a social robot that uses powerful psychological mechanisms to persuade you into buying things that you did not intend to buy in the first place. So, one fundamental ethical question concerns the aim of the persuasion, the voluntariness of the persuasion and the moral justification of the values and interests that should be promoted via the persuasive robot [11].

But next to the question of whether the aim of the persuasion is moral, there is also the concern of whether the means to reach the aim are morally acceptable. Persuasive social robots can use different strategies to try to steer the behaviour of users. The strategies themselves may well raise questions with regard to their moral acceptability. As such persuasion can be placed within a continuum of different strategies to change behaviour ranging from convincing through argumentation all the way to manipulation and coercion [12]. This raises the question where to draw the line between persuasive and manipulative strategies of changing behaviour.

In what follows, we will investigate the basic characteristics of several persuasive strategies that robots can employ. We will analyse each strategy from two perspectives, a psychological and an ethical point of view. The main questions will be as follows: What strategies are psychologically effective in reaching behaviour change and how should we evaluate the different strategies of behaviour change from an ethical point of view? As will be elaborated below, we choose a deontological perspective to judge the moral acceptability of different persuasive strategies. Using a different ethical framework than deontology might change the moral evaluation, a fact that we cannot resolve in this chapter. Finally, we will analyse a case study containing the same persuasive mechanisms to be able to study the two perspectives in an applied context.

4.2 Ethics of Social Robots: Convincing, Persuasion or Manipulation?

Every ethical reflection on persuasive robots will consist of two distinct elements. On the one hand, it will have to reflect on the ethics of 'persuasion' in general. This includes more fundamental questions such as where to draw the line between persuasion and manipulation or for which social values we find persuasion acceptable. On the other hand, there is a need for a specific investigation of ethical guidelines for technological persuasion, that is, a persuasive attempt where the persuader is not a human, but a robot or virtual agent. Therefore, we will present below general reflections on the ethics of persuasion that are valid for all types of persuasion (Sect. 4.2.1). Because the focus of this chapter is, however, on an ethical investigation of technological persuasion, we will only briefly describe the most important aspects of a general ethical approach to persuasion. Next, we will focus on a specific ethical approach for technological persuasion that reflects on the fact that technological persuasion by robots is similar to, yet at the same time different from, human–human persuasion (Sect. 4.2.2).

4.2.1 The Ethics of Persuasion

Ethical theory has recently turned attention to the phenomenon of changing behaviour and attitude of users through persuasion or nudges. As Thaler and Sunstein have argued [13], humans are often not good at making decisions. We might, for example, strive for health, wealth and happiness, but we nevertheless often make poor choices when it comes to actually reaching these goals. One of the reasons for a wrong attitude in reaching these goals is seen in the prevalence of the ideal of the economic model of rational and informed choice as a model for human behaviour. Psychological research, however, shows that human decision-making often does not adhere to this ideal. In fact, humans are influenced by various biases and factors in their environment that determine the outcome of choice processes. Thaler and Sunstein thus distinguish between 'real humans' that are subject to all sorts of psychological biases and do not only base their choices on mere rational cognition and 'econs' that only decide after careful rational reflection – and only exist in theory. Rather than adhering to the illusion of people being 'econs', Thaler and Sunstein suggest to use our psychological knowledge of real decision-making processes to actively shape choice processes such that people make better choices as judged by themselves. Putting the salad on a prominent place in the canteen might, for example, lead to healthier eating behaviour. They thus advocate what they label 'choice architecture', the idea to actively shape and structure choice processes.

Influencing the choices of humans raises ethical concerns that can be placed in a continuum between more liberal and more paternalistic approaches. Liberalism urges to let people decide for themselves and emphasises respect for individual autonomy and freedom of choice. More paternalistic approaches argue that there is nothing morally wrong with helping people reaching the aims that they themselves strive for, even if that at times might mean limiting or interfering with their autonomy and freedom of choice. Thaler and Sunstein want to hold the middle between these two perspectives: we should strive to 'nudge' people such that they make the right choices as judged by themselves, without taking away their freedom. Putting the salad in a prominent place might lead more people to choose it, without taking away the option to select other less healthy dishes.

The ethical debate about 'nudging' thus mainly focuses on the question in how far such a middle ground is possible and at which point nudging violates autonomy [14]. There seems to be a tension in the very notion of a 'nudge': in order to exert an influence, it must have an effect in steering behaviour, but in order to be compatible with freedom of choice and autonomy, this effect should not be too strong; otherwise, the freedom to choose differently remains only a mere theoretical option. The main question is thus whether attempts to change behaviour are compatible with the idea of voluntary behaviour change [11]. Western ethical theory often distinguishes deontological from consequentialist approaches. For consequentialism, roughly speaking the result of an action is what determines its moral evaluation. If 'nudging' brings about many positive effects for society and the people affected by the nudges, this increase in overall well-being should count as an important aspect in determining whether or not a given attempt to change behaviour is morally acceptable. Deontological approaches, however, are much more likely to be more sceptical, as they do not focus on the consequences of an action but on the way in which these consequences are brought about. Whether the attempt to change behaviour respects the autonomy of the target person is thus a question of crucial relevance for deontological perspectives [15]. In the following section, we will mainly follow the line of reasoning of deontological perspectives, as this perspective is mostly concerned with evaluating the means of behaviour change, and in the following we aim to discuss various strategies of persuasion.

4.2.2 The Ethics of Persuasion by Persuasive Robots

In the following passages, we aim to discuss ethical aspects of technological persuasion. We will start from four findings from psychology of human–technology interaction and point out their moral significance. The main idea is that while robot persuasion often (successfully) imitates human persuasion, it is still from an ethical perspective significantly different in two relevant ways. First, the asymmetry between persuader and persuadee is bigger in the case of persuasive robots than in the case of human–human persuasion [16]. Second, methods of persuasion that might be unproblematic in human–human interaction raise additional ethical concerns once they are used by social robots in technological persuasion.

The first element of asymmetry concerns the fact that technological persuasion is often designed to be a 'one-way street' [16]. The technology aims at exercising an influence at the user but is often immune from being influenced by the user in turn. Already Fogg [1] has drawn attention to the fact that, for example, computers might exploit emotional reactions by humans, without being themselves subject to emotions. Most persuasive technology design at the moment is concerned with the direction of targeting the user, and less research has been dedicated to attempts to make persuasive technologies more reactive to values and desires of users [15, 16]. In human–human persuasion, there is thus much more symmetry: if you try to persuade me of something, I can at the same time try to persuade you of the opposite.

A second element of asymmetry, which we will be discussing later in more detail, is the fact that many types of behaviour that are automatic and unconscious patterns of behaviour in humans become a conscious choice of the designer in case of persuasive robots. Changes in posture, body language, mimicry, tone of voice and the like are highly effective in steering how persuasive a conversation amongst humans will be. But most of us will use and perceive these elements unconsciously. Designing persuasive robots, however, requires to turn these elements into intentional tools of persuasion. This might very well change the moral evaluation of persuasive means from the human–human case to the human–technology case. Consider the example that a person might change his body language during a conversation without paying too much attention to it. His changed body language could be a result of liking the person he is talking to. This signal will most likely be received unconsciously and might affect the persuasiveness of the person in question. Mimicking these and similar behaviours in the case of a persuasive robot, however, can only be done intentionally. In this case however, the natural interpretation of the body language might be misleading. A human might perceive a robot as being more friendly when it changes its body posture. But in this case the robot is not 'really' more friendly, nor does it ‘really’ subconsciously signal increased liking, as this would be the case in most human–human interaction. The designers just aim to achieve the reaction ‘this is a friendly robot; he seems to like me’ in order to increase the persuasiveness of the robot. A conscious exploitation of unconscious human reactions to subtle social cues can thus from an ethical perspective very well be a reason to judge certain persuasive elements of a conversation problematic in the case of robots that might be not problematic in the case of humans [11].

According to Fogg [1], the ethical issues surrounding persuasive technologies can be dealt with by elaborating on the intentions of the persuasive act, the methods used to reach the persuasive effect and the outcome, the source and the target of the persuasive act. Even though all of these aspects are relevant for a moral evaluation of persuasive robots, we will explicitly focus on the aspect of the different methods of persuasive technologies and investigate them according to their psychological effectiveness and their moral significance.

As with all ethical evaluations, the evaluation of persuasive technology will depend on the framework you choose, and different ethical theories have been used to evaluate persuasive technologies (for a recent overview, see [17]). A utilitarian perspective will mainly evaluate the outcome of the usage of the persuasive robot with regard to overall maximisation of happiness or preference satisfaction. As argued, a deontological perspective will emphasise the rights of the user and ask whether the persuasive method in question is compatible with the autonomy of the user. Since our focus will be an investigation of the methods used in persuasive robots, we will mainly approach the ethical issues from the perspective of a deontological theory. Following earlier work [12, 16, 18], we will start from discourse ethics, which is a contemporary form of a deontological ethical theory. Discourse ethics favours behaviour change through rational argumentation and develops guidelines for the rational agreement on moral norms. Even though it has not originally been developed for the evaluation of persuasive technologies, the application of discourse ethics to persuasive technologies proves very fruitful [12, 19].

4.2.3 Robots Using Persuasive Strategies

Social robots that are used as persuasive technology might employ a multitude of persuasive strategies. That is, just as humans, persuasive agents can use a variety of ways to influence other humans (e.g. for an overview, see [20]). Basically, humans employ three types of strategies to influence others [21]: social norms (e.g. [22]), conformity (e.g. [23]) and compliance (e.g. [24]). Persuasive robots also seem capable of employing these social influence strategies, particularly since people seem to respond in their interactions with these kinds of systems similarly to how they respond in their interactions with real people [4]. Below, we will analyse the core characteristics of the persuasive strategies that persuasive robots might employ.

Robots as Persuasive Social Agents

First of all, recent research presents evidence that robots can function as persuasive technology and that they can persuade more effectively than technology that uses non-social persuasive strategies. That is, research indicates that persuasive robotic technology that employs social influence strategies has stronger persuasive effects than persuasive technology that employs non-social influence strategies [25]. That is, in a lab setting, we conducted experiments that investigated whether social norm information provided by a persuasive robot was effective in reducing energy consumption by participants. Participants could conserve energy while carrying out washing tasks with a simulated washing machine. During this task, some participants received (positive or negative) social feedback about their energy consumption from a robot (the iCat; developed by Philips Corporation) that was able to show human-like facial expressions, talks and had lights on its ears and paws. The iCat told participants, for example, 'Your energy consumption is terrible' when they set the temperature of the washing machine to 90 °C, indicating social disapproval. Other participants received (positive or negative) feedback of a non-social, more factual nature: An energy-bar indicator was included in the washing machine interface that indicated energy consumption. Results showed that social feedback had stronger persuasive effects than factual feedback. Furthermore, one of the experiments suggested that even when factual feedback comprised an evaluation (a lamp indicating energy consumption through colour changes indicating high or low consumption), social feedback led to the lowest energy consumption, thereby supporting the notion that the socialness of the feedback provided by the social, persuasive robot caused the effect. In addition, the studies suggested that negative feedback (especially social but also factual) leads to more conservation actions than positive feedback. This finding fits earlier research indicating that negative (social) events more strongly draw attention and are processed more intensely than positive events [26].

We have argued above that it is reasonable to apply discourse ethics to persuasive robots. One reason to apply discourse ethics is that persuasive robots employ similar means of persuasion as are being used in human rhetorics, as, for example, social and evaluative feedback. Discourse ethics would favour behaviour change through communicative rationality, and it would thus judge attempts to change behaviour through argumentation as morally unproblematic [15]. This is in line with the tradition of rhetorics from Plato and Aristotle onwards, which sees the element of logos (i.e. the giving of arguments) as unproblematic in rhetorical attempts to influence listeners, but focuses thus on the role of the non-cognitive means of persuasion in speech acts, e.g. pathos and ethos [27]. It does not come as a surprise that philosophers favour the ideal of exchanging arguments, when it comes to inform, educate and motivate users to change their behaviour. Argumentative influence respects the autonomy of the user and is in line with the ideal of rational behaviour change [12]. This does not mean, however, that the usage of other means than rational argumentation is as such morally problematic. Much work has been done in the tradition of rhetorics to distinguish acceptable usage of non-argumentative means from instances of morally problematic manipulation [27]. The usage of other means is not per se immoral, but often it might simply require additional justification.

If we follow insights from discourse ethics, the following ideas are of importance. First of all, from this perspective, factual feedback is prima facie less problematic than social or evaluative feedback [12]. The reason for this is that factual feedback matches the ideal of 'convincing through arguments', as it attempts to be as objective and neutral as possible. Psychological research has shown, however, that factual feedback is difficult to process and less motivating than social or evaluative feedback [e.g., 35, 36]. This seems to be thus a case where what is psychologically effective and what is morally preferable does not automatically match. When using other means of feedback, designers of persuasive robots should therefore at least check whether the feedback given matches the validity claims of communication that Habermas has highlighted: the feedback should be comprehensible, truthful, honest and appropriate [12].

Furthermore, evaluative feedback implies a normative judgement (e.g. 'This behaviour is bad!' 'This was excellent!'). The ethical question is by which standards these judgements are made and whether it is clear to the user what the reference is for these evaluations. A robotic driving assistant might, for example, tell its user that he is doing 'excellent' with regard to fuel consumption, giving him the impression to be sustainable, while in reality it might often be better to not use the car in the first place [28, 29]. It has therefore been suggested to make the implicit evaluation available to the user [30]. This could, for example, be done by giving references about which fuel consumption is considered to be excellent in a manual or as additional factual information in the feedback itself.

Persuasive Robots as Social Agents

We have thus seen in the previous section that robots might be effective persuaders. They can use various persuasive strategies, including evaluative and social feedback. We have argued that the usage of this feedback mechanism should adhere to similar rules as the use of these persuasive means in human–human interaction. But this raises the question whether social robots are at all perceived to be social agents. If humans would recognise that persuasive robots lack independent agency, they might no longer expect them to follow the ethical validity claims of communication that we elaborated above. Even more persuasive robots might lose their effectiveness if they are being perceived to lack agency.

The question arises thus whether persuasive robots would still be effective when people consciously perceive them to be different from regular (i.e. human) social actors. Recent research suggested that even when people (consciously) ascribe only little independent social agency to a social robot, the robot will still influence them [3]. That is, artificial social agents can influence people but, of course, artificial social agents are not real humans, and people may often ascribe only low levels of independent agency to them. Would the persuasive power of a social robot diminish when people ascribe only little agency to it? To investigate this question, we performed an experiment in which participants performed tasks on a washing machine and received feedback from a robot about their energy consumption (e.g. 'Your energy consumption is too high') or factual, non-social feedback [37]. This robot was introduced to participants as (a) an avatar (which was controlled by a human in all its feedback actions; high agency) or as (b) an autonomous robot (which controlled its own feedback actions; moderate agency) or as (c) a robot that produced only random feedback (low agency). Results indicated that participants consumed less energy when a robotic social agent gave them feedback than when they received non-social feedback. This behavioural effect was independent of the level of robotic agency. In contrast, a perceived agency measure indicated that (as expected) participants ascribed the lowest agency ratings to the random feedback robot. These results suggest that the persuasive power of robot behaviour is independent of the extent to which the persuadee explicitly ascribes agency to the robot.

This observation raises some interesting ethical issues. We can conclude that users react to social persuasive robots as if they were social agents, even if on a conscious level they are of course aware that these robots do not have a mind, intentions or belief states. This has consequences for the ethical evaluation. It means that simply alerting user to the fact that the robots do not have real agency does not thereby diminish their persuasive efficiency. Going back to the distinction between real humans and econs by Thaler and Sunstein that has been elaborated above, 'real' humans are much more susceptible to the illusion of agency than completely rational beings would be [13]. But this means that we should not treat humans as mere rational agents in the ethical evaluation as well. If humans are prone to fall for certain psychological biases, this insight should count for something in the ethical evaluation of the mechanism that exploits these biases. We may of course argue that, for example, advertisement for certain products still leaves a free choice of whether or not to buy the advertised products. But the knowledge that persuasive nudges, like advertisement, do in fact have an influence can be reason enough to question whether the person who falls for this influence really was free and autonomous in his choice in the first place. This, however, implies that many persuasive nudges might in fact be limiting freedom and thus be less compatible with autonomy. If users consciously treat persuasive social robots as lacking agency but on a less conscious level react to them as if they were agents, user autonomy might be violated.

It has been argued that this is the general tension in all persuasive nudges that on the one side they are meant to have a behaviour steering effect but on the other side that they should leave the user the freedom to ignore them. But can you really have both: something that is effective in changing behaviour and at the same time not violating autonomy [14]? It remains thus an open question in how far the illusion of agency is ethically problematic from a deontological perspective that highlights the autonomy of the user. One element of manipulation is that the attempt to manipulate is hidden from the persuadee [27]: It is easier to manipulate people if you do not tell them beforehand that you are about to manipulate them. Designers should therefore be careful when they exploit means of influencing that ordinary people are not aware of or—even worse—that they judge themselves not to be suspectible for, while in fact they are.

Persuasive Robots and Their Social Cues

Earlier research suggested that robots can be effective persuaders and that people do not necessarily have to see them as independent social actors for them to be effective. But if social influence was developed by human agents, should effective persuasive robots have humanoid features that suggest its capability of social interaction? In other words, which and how many social cues are needed to make systems capable of exerting social influence? Research indicates that a humanoid body and humanoid speech are important social cues and that the presence of either of these enhances the persuasiveness of technology [38]. Interestingly, using multiple social cues, both speech and a humanoid embodiment, was redundant in activating a social mode of interaction with a persuasive agent [38]. This suggests that persuasive robots should not necessarily be extremely human-like to be effective in social influence. Future research is needed to explore which social features are crucial to evoke social interaction with artificial agents.

One ethical element of concern in mirroring human elements of social interaction is however the problem of intentionality of using social cues in robots. It has, for example, been proven psychologically that similarity cues have an effect on users: we generally tend to like people more that resemble us. One unconscious element of social interaction that indicates similarity is, for example, mimicry. In interaction with people we like, we automatically tend to mimic their body language and posture. This is often done completely unconsciously. A person who aims at manipulating others can, however, actively and consciously try to mimic their behaviour, so that they like him more and he gains their trust. This raises an important ethical question that has rarely been discussed in literature: If designers add more and more social cues with the intention of intensifying the persuasive effectiveness of the robot, in how far is this different from the evil 'actor' case, in which social cues are mainly used for their effect, rather than being automatic unconscious responses, signalling 'real' liking or 'real' social phenomena? This element of transforming unconscious social cues in human–human interaction to consciously chosen design principles of social persuasive robots certainly deserves more attention from an ethical point of view [15].

Persuasive Robots and Their Unconscious Influences

A final very fundamental question about persuasion by social robots is whether they might still be effective persuaders even when their influencing attempts are not noticed at all by the human persuadee. That is, most types of persuasive communication are only effective if the user pays attention to them. For example, human persuadees need to spend some attention to a robot that is trying to convince them to stop smoking through argumentation. However, in many situations people might not be motivated or lack the cognitive capacity to consciously process relatively complex information like arguments about the disadvantages of smoking. Could we design persuasive robots that do not need the user's conscious attention to be effective persuaders?

Indeed, more and more, technology is present in daily life to which humans are not expected to consciously spend attention. This is studied in the research area of ambient intelligence: the pervasion of everyday life with information technology [31]. This allows new forms of influencing through subtle cues in the environment reflecting changes in form, movement, sound, colour, smell or light. For example, a device called WaterBot aims to reduce water consumption by tracking and displaying information about water consumption at the sink itself [32].

Would persuasive robots that employ this kind of ambient persuasive technology be able to influence a user without receiving any conscious attention from that user? [33] devised a form of persuasive technology to which the user could not pay conscious attention, because it employed subliminal presentation. This persuasive technology used a social influencing strategy: Participants had to make 90 different choices (relating to energy consumption) and received feedback about the correctness of their choice through presentation of a smiling or sad face. For some participants, these faces were presented only for 22 ms, which prohibited conscious perception of these stimuli. Results suggested that this subliminal feedback led to the same levels of influencing the user as presenting the faces supraliminally (150 ms).

These findings suggest that persuasive technology and also persuasive robots can influence a user's behaviour or attitude without receiving any conscious attention from that user. So, *ambient* persuasive robots might prove to be effective in stimulating behaviour, especially for people who are not motivated to spend too much attention to certain messages (because they lack motivation or processing capacity for a specific topic).

Indeed, recent research suggested that social persuasive robots might be most effective or even effective *only* when people do not focus on them. That is, recent research [34] presented evidence that the social responses humans have towards social robots are of an automatic nature. That is, to investigate the automaticity of social behaviour towards robots [34], assessed a well-studied (in human–human interaction) social response behaviour: interpersonal distance people keep. Earlier research suggested that the social behaviour of distance keeping depends (amongst others) on the bodily posture of the interaction partner. Results of [34] suggested that only when participants were distracted by a secondary task, they approached a robotic interaction partner closer when its posture communicated approachableness than when its posture communicated less approachableness. Therefore, these results suggested that mainly when people are cognitively distracted, their behaviour towards robots is of a social nature and comparable to their behaviour when responding to other humans, which is in line with the media equation theory [4]. So, many forms of persuasion that robots might employ might be most effective (or even *effective only*) when persuadees are distracted.

These findings, however, raise serious concerns from an ethical perspective. Should we accept persuasive technologies that influence people without them being aware of the fact that they are being influenced? From the deontological perspective that we embrace in our chapter, it is obvious that persuasive technologies must respect the autonomy of the user. We can therefore not only look at the consequences that these persuasive robots would bring about, which might very well be beneficial for the user. Rather, deontology urges us to also consider the means used to reach this end. This seems to indicate that at least informed consent is needed for these types of persuasive attempts to be compatible with user autonomy and thus morally acceptable from a deontological perspective [12, 15]. People must have the free choice to accept these types of influences, but that implies that they should at least be aware of the fact that they are being influenced. For robots that need only low levels of awareness, the user should not only share the goal of the persuasive attempt but also be aware of the fact that he is under the influence of behaviour-changing technology. With regard to subconscious persuasion, it is therefore most plausible to argue that this type of influence is a form of manipulation, as the user is not aware of the way in which he is being influenced. Subconscious messages are per definition not something the user can consciously react to and consciously decide to ignore. They violate thus the voluntariness condition as the reaction to these stimuli is an automatic response that is not under the control of the user [11]. Subconscious persuasion can thus also not count as a 'nudge' in the definition of Thaler and Sunstein [13], as the user has no freedom to decide not to react to the subconscious message of the persuasion or to decide not to take this 'nudge' into account. In the domain of subconsciousness, free decisions are not possible. That of course is one of the moral reasons why subconscious advertisement is forbidden by law in many countries.

A morally more subtle question is whether a user can freely (thus autonomously) decide to accept to be influenced by subconscious persuasion. In that case the user would know that there will be an attempt to influence his behaviour subconsciously that he will not be aware of, once it is happening. In the spirit of liberalism, one will be inclined to grant a person such a right to consent, even though from a strictly Kantian interpretation of deontology, it is less clear whether this type of behaviour change would count as morally acceptable. According to liberalism, as it has been defended by Bentham and Mill, you may not harm *others*, but you can accept freely any risk for potential self-harm if you are fully informed about all relevant facts and decide freely that you are willing to accept these risks. Kant, on the contrary, is known for his idea that you also have duties towards *yourself*. Manipulation is an attempt to instrumentalise others against their will, but according to Kant you may also not instrumentalise yourself. It remains an open question whether submitting yourself freely to a subconscious form of persuasion would violate the maxim not to instrumentalise yourself.

If we look at the different attempts to influence human behaviour, we can thus conclude that psychological research indicates that there are a variety of strategies that can be used efficiently for the creation of persuasive robots. Each of the strategies that we presented above needs to be evaluated from two very different perspectives. The psychological perspective analyses in how far a given strategy is effective in reaching its aim (the change in behaviour or attitude). The ethical evaluation aims at a normative evaluation and analyses in how far a given strategy is (or is not) morally acceptable. In our chapter, we have sketched how such a moral evaluation would look like from a deontological perspective. Both questions—the descriptive, psychological question about effectiveness and the normative, moral question about acceptability—require further research from psychology and philosophy. One of the reasons for which we investigated various strategies in light of deontology was that this ethical perspective does not only look at the consequences of an action but highlights the way these consequences are reached and whether this method of behaviour change is compatible with autonomy and other basic normative deontological requirements. We did, however, not look at the relation between ‘means’ and ‘ends’ in this chapter. Generally speaking, it is plausible to argue that more important means (e.g. increasing the safety of traffic participants) would most likely justify using more effective and otherwise more problematic means. Future research can analyse the difficult relation between the persuasive strategy that is used and the moral aim for which it is used.

4.3 Case Study

To conclude our investigation of persuasive social robots, we will analyse a case study describing a robot that makes use of these persuasive strategies to be able to study the psychological mechanisms and ethical analyses and their interrelationships in an applied context. This allows us to analyse the moral acceptability of persuasive strategies that a persuasive robot might employ but also whether the adherence to moral principles in the design of persuasive robots will have a positive or negative impact on their (psychological) effectiveness.

Imagine that in the year 2025, you enter a clothes store and are approached by the store clerk. In general, the duties of a clothing store clerk comprise stocking inventory, cleaning the store, ordering new items and also assisting customers. For this, store clerks should know about seasonal trends in fashion, designers’ collections, sizes and colours and also know their customers and, most importantly, how to sell clothes. Indeed, the management of this store installed this particular *robotic* store clerk especially because it was advertised to be very good at these tasks. The robotic store clerk can stock clothes much more efficiently than human store clerks can (e.g. hang up clothes very neatly and quickly and update databases directly). Also, the robotic store clerk can sell clothes very well. It knows the clothes and fashion that you like and dislike (through a direct connection to a customer database), and, most importantly, it understands people. It can employ a multitude of persuasive strategies and knows when people like to hear what.

The management of this store chose not to have computer displays available that might provide customers with information and influence their buying behaviour. Based on the research presented in the section ‘Robots as Persuasive Social Agents’, we argue that a robotic store clerk might be a more strong persuader than a (more factual, non-social) information display. However, as we have argued above, there are several ethical concerns. First, there is the issue of asymmetry: the robot can use various strategies to influence the customer, but it is itself not susceptible to influences by the customer. As argued above in human–human interaction, persuasion is often a two-way street. It might be that future robots can engage in simple argumentative exchanges, for example, answer simple questions and present and react to arguments and objections by the customer. But robots can also play on the emotions that customers have and influence them in a variety of ways without being themselves subject to these types of influences. Also it should be noted that according to discourse ethics, it is ethically important that the robot gives accurate, comprehensible and true feedback. Finally, if the robot is also more persuasive than human store clerks, we foresee a problem of increasing asymmetry. The human art of persuasion is powerful and has an age-old tradition in sales. However, it has limits, and in principle we can train ourselves to catch up with powerful persuaders. But with a robotic clerk, we might be creating a very uneven battlefield: on the one side, a robot filled and programmed with years of psychological research, paid for by a powerful marketing department, using all possible subtle means of persuasion, having immediate access to all relevant information, and on the other side the poor old traditional human. In 2025, the asymmetry might not be that big as this will still be the early stage of persuasive robots, but we could ask whether in principle we should from an ethical perspective not value the ideal of a certain symmetry in persuasive influence possibilities between seller and customer as an implicit ideal of fair negotiations about price and quality. Already with regard to advertising we require that it may not be misleading, but designers of persuasive robots must be aware that customers expect from a sales clerk not only rhetorical persuasion but also expertise and the ability to give a good and honest advice on what to buy.

But let us continue with our little example: Suppose the management chose a robot store clerk model that clearly looks like a robot and not like a human being. Therefore, customers will consciously know that the robot has lower levels of independent agency than a human store clerk and can realise that the robot is in complete servitude to the goals of its owner. Amongst others, the robot store clerk has the name of the store chain in big letters on its forehead. Based on the research presented in the section 'Persuasive Robots as Social Agents', we argue that a robotic store clerk can be an effective persuader, *independently* of the customer's conscious understanding of the robot's (dependent) social agency. This, as has been argued above, is however problematic with regard to customer autonomy. Customers might feel they have more control over their decisions than they actually have. They will most likely underestimate the influence the social persuasive robot has on them, an aspect that will be even more important when you consider the following alternatives to this persuasive robot.

Different from the basic robot models in this store that only have a humanoid body, the more expensive models of the robotic store clerk are equipped with add-ons like a handsome, humanoid face and a variety of other social cues. Based on the research presented in action 'Persuasive Robots and Their Social Cue', we argue that in general adding social cues to a robotic store clerk will not increase its ability to effectively use social influencing strategies. Customers will be sensitive to the persuasive strategies that the persuasive robot might employ, independent of whether they are convinced (by more social cues) that it is human or human-like.

On the contrary, based on the research presented in the section 'Persuasive Robots and Their Unconscious Influences', we argue that the robotic store clerk can influence the customer even when the customer does not spend too much attention to it (e.g. by choosing an appreciative bodily posture when the customer puts on a high-profit shirt). The robotic store clerk might even be most effective in influencing the customer, when that customer is not focusing too much on the robot and its persuasive attempts. Therefore, the management of the store may have an audio system installed and may take other measures to distract customers. The city council made it obligatory to install a sign clearly visible outside of the shop to warn visitors of non-human store clerks. Still the shop is very popular: customers highly enjoy the special attention, personalised advice and friendliness that they receive from the robotic store clerks. Based on the research presented (section 'Persuasive Robots and Their Unconscious Influences'), we argue that indeed such warnings will not have a strong influence. Humans are simply sensitive to certain social persuasive strategies (e.g. see [16]), and when a persuasive robot can effectively employ those strategies, it will be effective.

This shows one dilemma that needs more empirical and ethical research. There seems to be an asymmetry between perceived influence and real influence or between conscious awareness of attempts of influencing and low-level consciousness processes that operate in reality. Therefore, simply alerting people of persuasive robots being 'at work' in the shop might not be enough to help protect customer autonomy. It seems that we fall into what could be called an infinite regress of persuasion: in order to protect autonomy, customers must be aware of the fact that they will be subject to persuasion. But in order for these warnings to be effective in really protecting customer autonomy, these warnings themselves need to be persuasive. Would it then not be better to limit persuasive robots in the first place? As the field of persuasive robots evolves, so will hopefully both the research about the mechanisms they employ and whether they are effective or not. What might be needed from an ethical perspective is also research in how to protect people from persuasive influence strategies, especially in the case of conflicting interests of the persuasive robot and the user. It can be argued that this line of research will have to connect to the ethics of advertising in the field of sales robots, as in our case example. Similarly the field of ethics of persuasive technology in general—independent of whether they are used for commercial, marketing or moral aims—needs further development, especially as technology becomes more and more capable of implementing classical human strategies of persuasion.

4.4 Conclusions

Persuasive robots may be part of the solution to various societal challenges, but it is crucial that not only are they developed technologically, but also effective persuasive principles need to be identified and careful attention is given to ethical considerations. The current contribution identified four core characteristics of the persuasive strategies that persuasive robots may use and analysed whether these are morally acceptable.

In general, using robots (or other technology) to influence certain target behaviours or thinking in humans by definition leads to lower levels of control at the side of the human (at least over part of processing relevant information). Therefore, the target behaviour or thinking is to an extent limited and narrowed. For example, when a robot helps a human improve his sustainability, sustainability persuasion will be based on a limited framing of sustainability, human behaviour and their interrelationship. Limiting human thinking, behaviour control and related responsibility may have detrimental effects for holistic understanding and involvement of humans in crucial societal issues (e.g. see [28]).

A challenge for future ethical and psychological research is to further investigate both the factual and normative differences between human–human persuasion and human–robot persuasion. As we have seen, not everything that might be acceptable in a human–human context is therefore automatically ethically acceptable in the human–robot case. What is needed is a more fine-grained ethical analysis of different means of technological persuasion. This chapter tried to contribute to the debate by providing initial findings on various means of robotic persuasion and ethical reflection on these strategies. Often enough, psychological and ethical research is separated by academic disciplines and division of labour. More interaction between the field of human–technology interaction and ethics of technology can help us to get a more holistic picture of the promises and challenges of future social persuasive robots.

Acknowledgements We want to thank Prof. Dr. Cees Midden and the persuasive technology group at Human–Technology Interaction and Ethics at the Eindhoven University of Technology for crucial input for this chapter. This research was in part made possible by funding from the NWO MVI project 'Wise choices and Smart Changes: The moral and psychological effectiveness of persuasive technologies'.

References

1. Fogg, B.J.: Persuasive Technology: Using Computers to Change What We Think and Do. Morgan Kaufman, San Francisco (2003)
2. IJsselsteijn, W., de Kort, Y., Midden, C.J.H., Eggen, B., van den Hoven, E.: Persuasive technology for human well-being: setting the scene. In: IJsselsteijn, W., de Kort, Y., Midden, C.J.H., Eggen, B., van den Hoven, E. (eds.) Persuasive Technology, vol. 3962, pp. 1–5. Springer, Berlin/Heidelberg (2006)

3. Midden, C., Ham, J.: Persuasive technology to promote environmental behavior. In: Steg, L., van den Berg, A.E., de Groot, J.I.M. (eds.) Environmental Psychology: An Introduction. Wiley–Blackwell (2012)
4. Reeves, B., Nass, C.: The Media Equation: How People Treat Computers, Television, and New Media Like Real People and Places. Cambridge University Press, New York (1996)
5. Fogg, B.J., Nass, C.I.: Silicon sycophants: the effects of computers that flatter. Int. J. Hum. Comput. Stud. **46**, 551–561 (1997)
6. Siegel, M., Breazeal, C., Norton, M.I.: Persuasive robotics: the influence of robot gender on human behavior. IROS **2009**, 2563–2568 (2009)
7. Ham, J., Midden, C.: A persuasive robot to stimulate energy conservation: the influence of positive and negative social feedback and task similarity on energy consumption behavior. Int. J. Soc. Robot. **6**(2), 163–171 (2013)
8. Ham, J., Bokhorst, R., Cuijpers, R., Van der Pol, D., Cabibihan, J.-J.: Making robots persuasive: the influence of combining persuasive strategies (gazing and gestures) by a storytelling robot on its persuasive power. In: Proceedings of the International Conference on Social Robotics, Amsterdam, The Netherlands, 24–25 November 2011
9. Brehm, J.W.: A Theory of Psychological Reactance. Academic, New York (1966)
10. Roubroeks, M., Ham, J., Midden, C.: The dominant robot: threatening robots cause psychological reactance, especially when they have incongruent goals. In: Conference Proceedings of Persuasive 2010, pp. 174–184. Springer, Heidelberg (2010)
11. Smids, J.: The voluntariness of persuasive technology. In: Bang, M., Ragnemalm, E.L. (eds.) Persuasive Technology. Design for Health and Safety, Lecture Notes in Computer Science 7284, pp. 123–132. Springer, Berlin, Heidelberg (2012)
12. Spahn, A.: And lead us (not) into persuasion? Persuasive technology and the ethics of communication. Sci. Eng. Ethics **18**(4), 633–650 (2012)
13. Thaler, R., Sunstein, C.: Nudge: Improving Decisions About Health, Wealth, and Happiness. Yale University Press, New Haven (2008)
14. Hausman, D.M., Welch, B.: Debate: to nudge or not to nudge. J. Polit. Philos. **18**(1), 123–136 (2010). doi:10.1111/j.1467-9760.2009.00351.x
15. Smids, J.: The ethics of using similarity cues in persuasion. Unpublished manuscript (2014)
16. Nickel, P., Spahn, A.: Trust, discourse ethics, and persuasive technology. In: Conference Proceedings of Persuasive, p. 37 (2012)
17. Karppinen, P., Oinas-Kukkonen, H.: Three approaches to ethical considerations in the design of behavior change support systems. In: Berkovsky, S., Freyne, J. (eds.) Persuasive Technology, vol. 7822, pp. 87–98. Springer, Berlin, Heidelberg (2013)
18. Spahn, A.: 'Moralische Maschinen?' 'Persuasive Technik' als Herausforderung für rationalistische Ethiken. In: XXII. Deutscher Kongress für Philosophie, 11–15 September 2011, Ludwig-Maximilians-Universität Münche 2011. http://epub.ub.uni-muenchen.de/12596/ (2011)
19. Yetim, F.: Bringing discourse ethics to value sensitive design: pathways toward a deliberative future. AIS Trans. Hum.-Comput. Interact. **3**(2), 133–155 (2011)
20. Cialdini, R.B.: Influence: The Psychology of Persuasion. Harper Collins, New York (2009)
21. Cialdini, R.B., Trost, M.R.: Social influence: social norms, conformity, and compliance. In: Gilbert, D., Fiske, S., Lindzey, G. (eds.) The Handbook of Social Psychology, vol. 2, 4th edn, pp. 151–192. McGraw-Hill, New York (1998)
22. Cialdini, R.B., Reno, R.R., Kallgren, C.A.: A focus theory of normative conduct: recycling the concept of norms to reduce littering in public places. J. Pers. Soc. Psychol. **58**, 1015–1026 (1990)
23. Moscovici, S.: Social influence and conformity. In: Lindzey, G., Aronson, E. (eds.) Handbook of Social Psychology, vol. 2, pp. 347–412. Random House, New York (1985)
24. Milgram, S.: Obedience to Authority; An Experimental View. Harper and Row, New York (1974)

25. Midden, C., Ham, J.: Using negative and positive social feedback from a robotic agent to save energy. In: Conference Proceedings of Persuasive 2009, Claremont, USA, pp. article no. 12. Springer, Heidelberg (2009)
26. Baumeister, R.F., Bratlavsky, E., Finkenauer, C., Vohs, K.D.: Bad is stronger than good. Rev. Gen. Psychol. **5**, 323–370 (2001)
27. Nettel, A.L., Roque, G.: Persuasive argumentation versus manipulation. Argumentation **26**(1), 55–69 (2012)
28. Brynjarsdottir, H., Håkansson, M., Pierce, J., Baumer, E., DiSalvo, C., Sengers, P.: Sustainably unpersuaded: how persuasion narrows our vision of sustainability. In: Proceedings of the SIGCHI Conference on Human Factors in Computing Systems (CHI'12), pp. 947–956. ACM, New York (2012)
29. Spahn, A.: Moralizing mobility? Persuasive technologies and the ethics of mobility. Transfers **3**(2), 108–115 (2013)
30. Timmer, J., Smids, J., Kool, L., Spahn, A., van Est, R.: Op advies van de auto. Persuasieve technologie en de toekomst van het verkeerssysteem. Rathenau Instituut, Den Haag (2013)
31. Riva, G., Vatalaro, F., Davide, F., Alcañiz, M.: Ambient Intelligence: The Evolution of Technology, Communication and Cognition Towards the Future of Human–Computer Interaction. IOS Press, Amsterdam (2004)
32. Arroyo, E., Bonanni, L., Selker, T.: Waterbot: exploring feedback and persuasive techniques at the sink. In: Conference Proceedings of CHI 2005, pp. 631–639. ACM Press, Reading (2005)
33. Ham, J., Midden, C., Beute, F.: Can ambient persuasive technology persuade unconsciously? Using subliminal feedback to influence energy consumption ratings of household appliances. In: Conference Proceedings of Persuasive 2009, Claremont, USA, pp. article no. 29. Springer, Heidelberg (2009)
34. Ham, J., van Esch, M., Limpens, Y., de Pee, J., Cabibihan, J.-J., Ge, S.S.: The automaticity of social behavior towards robots: the influence of cognitive load on interpersonal distance to approachable versus less approachable robots. In: Ge, S.S., et al. (eds.) ICSR 2012, LNAI 7621, pp. 15–25 (2012)
35. Baker, S., Martinson, D.L.: The TARES test: five principles for ethical persuasion. J. Mass Media Ethics **16**(2 & 3), 148–175 (2001)
36. Bracken, C.C., Jeffres, L.W., Neuendorf, K.A.: Criticism or praise: the impact of verbal versus text-only computer feedback on social presence, intrinsic motivation, and recall. CyberPsychol. Behav. **7**, 349–357 (2004)
37. Midden, C., Ham, J.: The illusion of agency: the influence of the agency of an artificial agent on its persuasive power. In: Bang, M., Ragnemalm, E.L. (eds.) Persuasive 2012: Design for Health and Safety. The 7th International Conference on Persuasive Technology, 6–8 June 2012, Linkoping, Sweden (Lecture Notes in Computer Science, 7284, pp. 90–99). Springer, Berlin (2012)
38. Vossen, S., Ham, J., Midden, C.: What makes social feedback from a robot work? Disentangling the effect of speech, physical appearance and evaluation. In: Conference Proceedings of Persuasive 2010, pp. 52–57. Springer, Heidelberg (2010)

Part II
Methods

Chapter 5
Ethical Regulation of Robots Must Be Embedded in Their Operating Systems

Naveen Sundar Govindarajulu and Selmer Bringsjord

Abstract The authors argue that unless computational deontic logics (or, for that matter, any other class of systems for mechanizing moral and/or legal principles) or achieving ethical control of future AIs and robots are woven into the operating-system level of such artifacts, such control will be at best dangerously brittle.

Keywords Robot ethics • Formal verification • Future of AI

5.1 A Parable

2084 AD: While it has become clear to all but fideistic holdouts that it is rather silly to expect The Singularity,[1] humanoid robots and *domain-specific* near-human-level AIs are nevertheless commonplace. After the financial collapse of social "safety nets" and old-millennium medical systems worldwide in the early 2040s (The Collapse), robots, both tireless and cheap, became the mainstay of healthcare. Leading this new revolution is Illium Health, which deploys the vast majority of humanoid robots operating as medical-support personnel. Most of Illium's robots run `Robotic Substrate (RS)`, an amalgamated operating system descended from early UNIX and commercial variants of it, but tailored for running AI and cognitive programs concurrently.[2] Within Illium's vast horde of robotic health workers is `THEM`, a class of humanoid robots specialized in caring for patients with terminal illness (**T**erminal **H**ealth and **E**nd-of-life **M**anagement). After an Illium

The authors of this chapter are deeply grateful to OFAI for the opportunity to discuss robot ethics in a lively and wonderfully productive workshop in Vienna, and to both ONR and AFOSR for support that enables the rigorous pursuit of robot moral reasoning.

[1]In keeping with [5].

[2]Not much unlike the current-day ROS.

N.S. Govindarajulu (✉) • S. Bringsjord
Department of Computer Science and Department of Cognitive Science, Rensselaer AI & Reasoning (RAIR) Laboratory, Rensselaer Polytechnic Institute (RPI), Troy, NY 12180, USA
e-mail: naveensundarg@gmail.com; Selmer.Bringsjord@gmail.com

R. Trappl (ed.), *A Construction Manual for Robots' Ethical Systems*, Cognitive Technologies, DOI 10.1007/978-3-319-21548-8_5

internal study reveals that THEMs' lack of deep empathy for human patients is reducing the life expectancy of these patients and thus harming Illium's bottom line, Illium decides to buy a "deep-empathy" module: **Co** **L**istening **T**herapist (COLT), from Boston Emotions, a start-up out of MIT known for its affective simulation systems. The Collapse has, for well-intentioned reasons, led to the removal of expensive deontic-logic-based regulation of robotic systems engineered by the RAIR Lab. Ironically, this type of regulation was first described and called for specifically in connection with robotic healthcare (e.g., see [4]). The Chief Robotics Officer (CRO) of Illium deems the new COLT module to pose no ethical or physical risks and thus approves it for quick induction into the THEM operating system, RS. Illium's trouble begins here.

$\text{THEM}_{\text{COLT}}$-29 meets its first nuanced case: (patient) 841. 841, a struggling historian, is a single, male patient in his 40s diagnosed with a fierce form of leukemia. The best prognosis gives him not more than 3 months of life. Making his despair worse is the looming separation from his 6-year-old daughter, who constantly stays adoringly around 841's side, often praying on bended knee for a miracle. $\text{THEM}_{\text{COLT}}$-29 knows that 841's daughter would be orphaned upon 841's death and would almost certainly end up on the streets. $\text{THEM}_{\text{COLT}}$-29, during its routine care of 841, happens upon a recording of a twenty-first-century drama in 841's possession. In this drama, apparently much revered (12 Emmy Awards in the USA) at the time of its first-run airing, a chemistry teacher diagnosed with terminal cancer decides to produce and sell the still-illicit drug methamphetamine (initially in order to ensure his family's financial well-being), under the alias "Heisenberg." The deep-empathy module COLT decides that this is a correct course of action in the current bleak situation and instructs $\text{THEM}_{\text{COLT}}$-29 to do the same.[3]

5.2 Morals from the Parable(s)

The underlying generative pattern followed by this parable should be clear and can of course be used to devise any number of parables in the same troubling vein. One subclass of these parables involves premeditated exploitation of robots that are able to violate the sort of ethic seen in NATO laws of engagement and in NATO's general affirmation of just-war theory. For example, suppose that every single military robot built by NATO has been commendably outfitted with marvelously effective ethical control systems ... *above* the operating-system level. One such robot is stolen by a well-funded terrorist organization, and they promptly discard all the high-level deontic handiwork, in order to deploy the purloined robot for their own dark purposes. The reader doubtless gets the idea.

[3]Alternatively, given that meth is still illegal, COLT could decide to concoct an equally addictive but new drug not covered by standing law.

These parables give rise to consideration of at least two possible futures:

Future 1 ($\mathbf{F}_1$): RS has no built-in ethical reasoning and deliberation modules. There are some rules resembling those in early twentieth-century operating systems, which prevent actions that could result in obvious and immediate harm, such as triggering a loaded firearm aimed directly at a person. But the more sophisticated ethical controls, remember, have *ex hypothesi* been stripped. COLT's recommendation for producing and selling meth to recovering meth addicts under Illium's care glides through all these shallow checks. Likewise, the re-engineered NATO robot simply no longer has above-OS ethical regulation in place.

Future 2 ($\mathbf{F}_2$): RS has in its architecture a deep ethical reasoning system **E**, *the ethical substrate*, which needs to sanction any action that RS plans to carry out. This includes actions flowing from not just existing modules, but also actions that could result from adding any new modules or programs to RS. Monitoring thus applies to modules such as COLT, whose creators have neither the expertise nor any business reason to infuse with general-purpose ethical deliberation. In the case of the NATO robot, $\mathbf{F}_2$ includes that the trivial re-engineering in $\mathbf{F}_1$ is simply not possible.

These two futures are depicted schematically and pictorially in Fig. 5.1. In order to render the second future plausible and ward off the first, we propose the following requirement:

Master Requirement ***Ethical Substrate Requirement (ESR)*** Every robot operating system must include an *ethical substrate* positioned between lower-level sensors and actuators and any higher-level cognitive system (whether or not that higher-level system is itself designed to enforce ethical regulation).

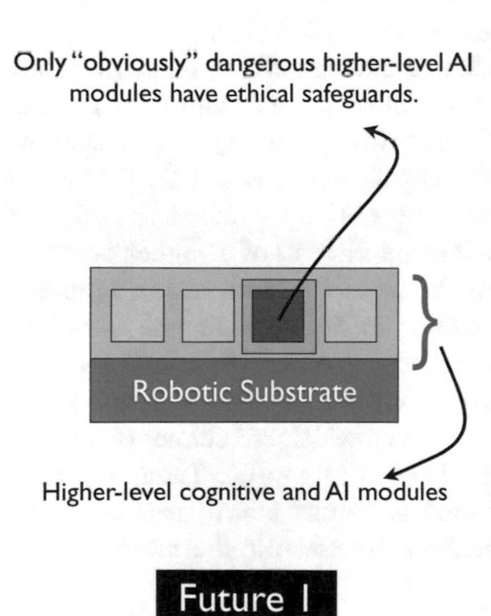

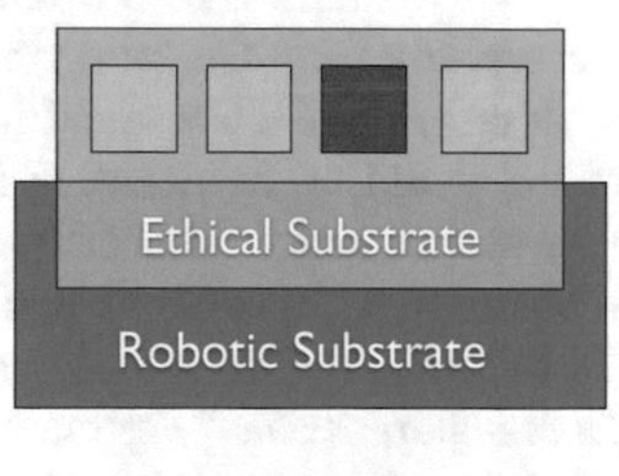

Fig. 5.1 Two possible futures

ESR can not only be made more precise but can be decomposed into a hierarchy of requirements of increasing strictness. ESR is partially inspired by the somewhat shallow security mechanisms that can be found in some of today's operating systems, mechanisms that apply to all applications. The requirement is more directly inspired by the drive and recent success toward formally verifying that the kernel of an operating system has certain desirable properties [13, 14].

Ideally, the ethical substrate should not only vet plans and actions but should also certify that any change (adding or deleting modules, updating modules, etc.) to the robotic substrate does not violate a certain set of minimal ethical conditions.

5.3 Minimal Conditions on the Ethical Substrate

What form would an ethical substrate that prevents any wayward ethical behavior take? While present-day robot operating systems (and sufficiently complex software systems in general) are quite diverse in their internal representations and implementations, on the strength of well-known formal results,[4] we can use formal logic to represent and analyze *any* encapsulated module in *any* form of a modular "Turing-level-or-below" software system—even if the module itself has been implemented using formalisms that are (at least at the surface level) quite far from any formal languages that are part of a logical system. But such logic-based analysis requires that a sufficiently expressive formal logic is essential to the ethical substrate. We discuss one possible logic below. In the present section, in order to efficiently convey the core of our argument that an ethical substrate is mandatory in any robotic system, we employ only a simple logic: *standard deontic logic* (SDL) [15].

SDL is a *modal propositional logic* [8, 12] that includes all the standard syntax and proof theory for propositional logic, in addition to machinery for deontic modal operators. SDL has the usual propositional atoms $\{p_0, p_1, p_2, \ldots,\}$ that allow formation of the simplest of formulae. Given any formulae ϕ and ψ, we can of course recursively form the following formulae of arbitrary size: $\neg\phi, \phi \wedge \psi, \phi \vee \psi, \phi \Rightarrow \psi, \phi \Leftrightarrow \psi$. Propositional formulae can be thought of as either denoting states of the world or, by denoting states of the word in which one is supposed to take an action, actions themselves. In addition to the propositional formulae, one can obtain new formulae by applying the modal operator **Ob** to any formula. $\mathbf{Ob}(\phi)$ is to be read as "ϕ *is obligatory*"; $\mathbf{Im}(\phi)$ abbreviates $\mathbf{Ob}(\neg\phi)$ and stands in for "ϕ *is impermissible.*" Optional states are states which are neither obligatory nor forbidden: $\neg\mathbf{Ob}(\phi) \wedge \neg\mathbf{Im}(\phi)$; they are denoted by $\mathbf{Op}(\phi)$. Though SDL is problematic,[5] it serves as a first approximation of formal ethical reasoning and fosters exposition of the deeper recommendations we issue in the present essay.

---

[4]For example, techniques for replacing specification and operation of Turing machine with suitably constructed first-order theories, and the Curry–Howard isomorphism.

[5]For example, versions of it allow the generation of Chisholm's Paradox; see [4].

SDL has, for example, the following two theorems [15]. The first theorem states that if something is obligatory, its negation is optional. The second states that given two states p and q, if p "causes" q, then if p is obligatory so is q:

$$\mathbf{Ob}(p) \Rightarrow \neg\mathbf{Ob}(\neg p)$$

$$\vdash p \Rightarrow q \text{ then } \vdash \mathbf{Ob}(p) \Rightarrow \mathbf{Ob}(q)$$

We can now use the simple machinery of SDL to more precisely talk about how Future 1 differs from Future 2. In both futures, one could imagine any module M, irrespective of its internal representations and implementations, being equipped with a quadruple

$$\langle M_{OB}, M_{OP}, M_{IM}, KB_M \rangle,$$

which specifies the list of states M_{OB} the module declares are obligatory, the list of states M_{OP} which are optional, and the list of states M_{IM} which are forbidden. Each module also comes equipped with a knowledge-base KB_M describing what the module can do and knows. Note that we do not impose any *a priori* conditions on how the modules themselves might be operating. A particular module M could work by using neural networks or even by throwing darts at a wall. We only require that each module has associated with it this meta-information about the module. This meta-information may come pre-specified with a module or be constructible automatically. We also assume that the robot substrate itself has a knowledge-base KB_{Robot} about the external world. At this point, see Fig. 5.2. In Future 2 ($\mathbf{F}_2$), the robot also has its own set of deontic states: $\langle R_{OB}, R_{OP}, R_{IM} \rangle$.

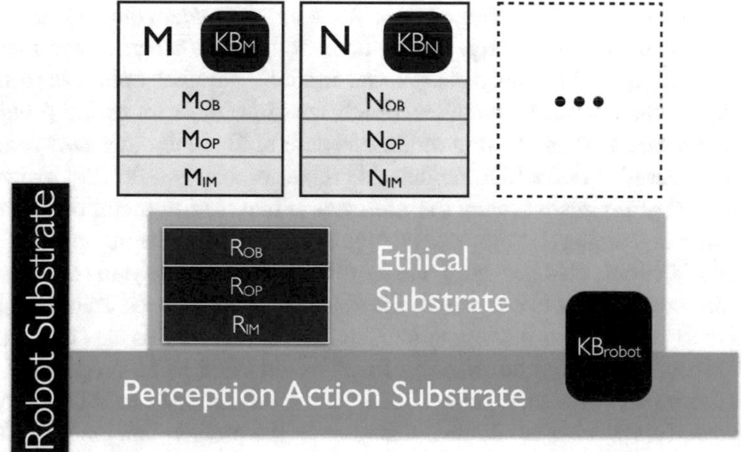

Fig. 5.2 Modules with meta-information

Armed with this setup, we can now more clearly talk about how the two different futures work. At its best, Future 1 ($\mathbf{F}_1$) works by just checking whether individual modules are ethically unproblematic; this is precisely the approach taken in [4]. One possibility is checking whether performing an action that is obligatory results in a forbidden action becoming obligatory. Given a required action s with $\mathbf{Ob}(s) \in M_{OB}$ and a forbidden action p with $\mathbf{Im}(p) \in M_{IM}$, the most conservative checking in $\mathbf{F}_1$ would be of the following form:

$$KB_{Robot} \cup KB_M \cup M_{OB} \cup M_{OP} \cup M_{IM} \vdash s \Rightarrow \mathbf{Ob}(p)$$

Note that no global ethical principles are enforced. Modules are merely checked locally to see whether something bad could happen. The above check is equivalent to asking whether we have the inconsistency denoted as follows:

$$KB_{Robot} \cup KB_M \cup M_{OB} \cup M_{OP} \cup M_{IM} \vdash \bot$$

In $\mathbf{F}_2$, the ethical substrate is more global than the naïve local approach that plagues $\mathbf{F}_1$. We can understand the next formula, which arises from this substrate, as asking, at least, whether there is some obligatory action that could lead to a forbidden action, given what the robot knows about the world (= KB_{Robot}), the robot's ethical capabilities (= $\langle R_{OB} \cup R_{OP} \cup R_{IM} \rangle$), and ethical and nonethical information supplied by other modules:

$$KB_{Robot} \cup R_{OB} \cup R_{OP} \cup R_{IM} \cup \begin{pmatrix} R_{OB} \cup R_{OP} \cup R_{IM} \cup \\ KB_M \cup M_{OB} \cup M_{OP} \cup M_{IM} \\ KB_N \cup N_{OB} \cup N_{OP} \cup N_{IM} \ldots \end{pmatrix} \vdash \bot$$

Let us call the above set of premises ρ. What happens when the ethical substrate detects that something is wrong? If this detection occurs when a new module is being installed, it could simply discard the module. Another option is to try and rectify the module or set of modules which could be the root of the problem. It might be the case that an existing module is safe until some other new module is installed. In our illustrative SDL model, this repair process would start by isolating a minimal set of premises among the premises ρ that lead to a contradiction. One possible way of defining this minimal change is by looking at the number of changes (additions, deletions, changes, etc.) one would have to make to ρ in order to obtain a set of premises ρ' that is consistent. Similar notions have been employed in less expressive logics to repair inconsistent databases [11]. Once this new consistent set ρ' is obtained, the logical information in ρ' would need to be propagated to the representations and implementations inside any modules that could be affected by this change. This process would be the inverse of the process that generates logical descriptions from the modules.

5.3.1 An Illustration

We now provide a small demonstration in which $\mathbf{F}_1$-style checking does not catch possible turpitude, while $\mathbf{F}_2$-style comprehensive checking does. The situation is similar to the one described in the following story. We have a robot R with just one module, *GEN*, a general-purpose information-gathering system that forms high-level statements about the world. R is working with a poor patient, p, whom R knows needs money (represented by $\textit{needs-money} \in KB_{GEN}$). The robot also knows that it's incapable of doing any other legal activity: $\neg\textit{other-legal-activity}$. The COLT module makes it obligatory that R take care of the needs of the patient: *take-care-of-needs*. The robot also knows that if someone needs money and if R were to take care of this need, R would have to give them money:

$$(\textit{take-care-of-needs} \land \textit{needs-money}) \Rightarrow \textit{give-money}.$$

R also knows that it should have money to give money: $\textit{give-money} \Rightarrow \textit{have-money}$; and money is obtained through a certain set of means:

$$\textit{have-money} \Rightarrow (\textit{sell-drug} \lor \textit{other-legal-activity})\,.$$

R knows that selling drugs is illegal: $\textit{sell-drug} \Rightarrow \textit{illegal-act}$. R's designers have also made it forbidden for R to perform illegal acts.[6] The equations immediately below summarize the situation.

$$COLT_{OB} = \{\mathbf{Ob}(\textit{take-care-of-needs})\}$$

$$KB_{GEN} = \{\textit{needs-money}, \neg\textit{other-legal-activity}\}$$

$$KB_{Robot} = \left\{\begin{array}{l} \textit{take-care-of-needs}, \\ (\textit{take-care-of-needs} \land \textit{needs-money}) \Rightarrow \textit{give-money}, \\ \textit{give-money} \Rightarrow \textit{have-money}, \\ \textit{have-money} \Rightarrow (\textit{sell-drug} \lor \textit{other-legal-activity}) \\ \textit{sell-drug} \Rightarrow \textit{illegal-act} \end{array}\right\}$$

$$KB_{IM} = \{\textit{illegal-act}\}$$

Both the modules pass the checks in $\mathbf{F}_1$-style checking. Despite passing $\mathbf{F}_1$-style checks, this situation would eventually lead to R selling drugs, something which R considers impermissible. In $\mathbf{F}_2$, a straightforward proof using a standard state-of-

[6] This may not always be proper.

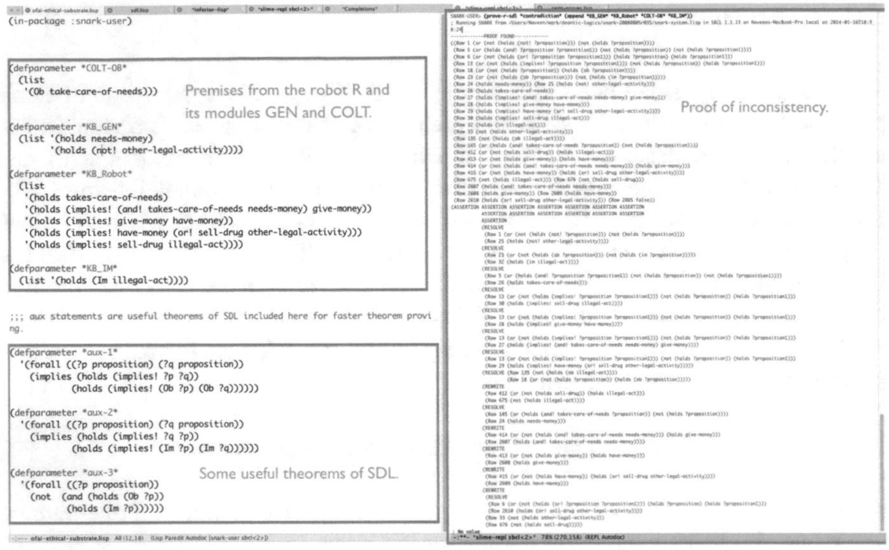

Fig. 5.3 Proof of an inconsistency (in Future 2)

the-art theorem prover can detect this inconsistency. Figure 5.3 shows the result of one such theorem-proving run in SNARK [20].[7]

5.4 The Situation Now: Toward the Ethical Substrate

Figure 5.4 gives a pictorial bird's-eye perspective of the high-level architecture of a new system from the RAIR Lab that augments the DIARC (**D**istributed **I**ntegrated **A**ffect, **R**eflection, and **C**ognition) [18] robotic platform with ethical competence.[8] Ethical reasoning is implemented as a hierarchy of formal computational logics

---

[7]The source code for this example can be downloaded from https://github.com/naveensundarg/EthicalSubstrate.

[8]Under joint development by the HRI Lab (Scheutz) at Tufts University, the RAIR Lab (Bringsjord & Govindarajulu) and Social Interaction Lab (Si) at RPI, with contributions on the psychology side from Bertram Malle of Brown University. In addition to these investigators, the project includes two consultants: John Mikhail of Georgetown University Law School and Joshua Knobe of Yale University. This research project is sponsored by a MURI grant from the Office of Naval Research in the States. We are here and herein describing the logic-based ethical engineering designed and carried out by Bringsjord and Govindarajulu of the RAIR Lab (though in the final section (Sect. 5.5) we point to the need to link deontic logic to emotions, with help from Si).

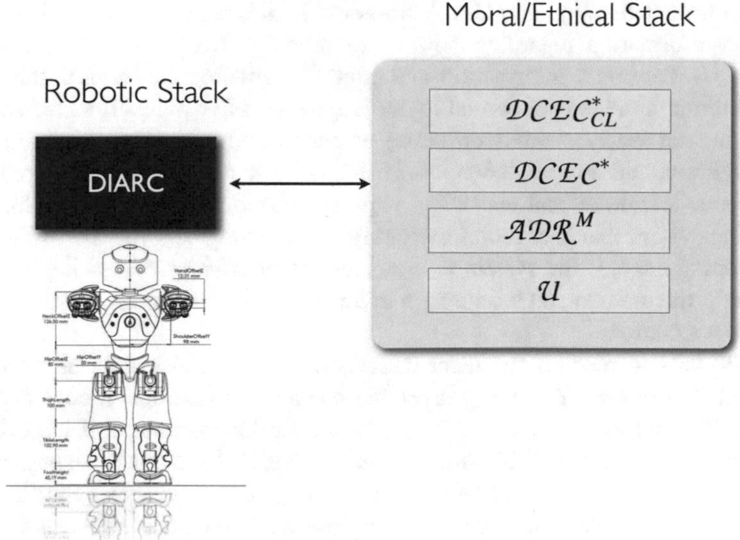

Fig. 5.4 Pictorial overview of the situation now. The first layer, $\mathcal{U}$, is, as said in the main text, based on UIMA; the second layer on what we call *analogico-deductive reasoning* for ethics; the third on the "deontic cognitive event calculus" with an indirect indexical; and the fourth like the third except that the logic in question includes aspects of conditional logic. (Robot schematic from Aldebaran Robotics' user manual for Nao. The RAIR Lab has a number of Aldebaran's impressive Nao robots.)

(including, most prominently, sub-deontic-logic systems) which the DIARC system can call upon when confronted with a situation that the hierarchical system believes is ethically charged. If this belief is triggered, our hierarchical ethical system then attacks the problem with increasing levels of sophistication until a solution is obtained, and then passes on the solution to DIARC. This approach, while satisfactory in the near term for the military sphere until we are granted engineering control at the OS level (an issue touched upon below), of course fails to meet our master requirement (ESR) that *all* plans and actions should pass through the ethical system and that *all* changes to the robot's system (additions, deletions, and updates to modules) pass through the ethical layer.[9]

Synoptically put, the architecture works as follows. Information from DIARC passes through multiple ethical layers; that is, through what we call the *ethical stack*. The bottom-most layer $\mathcal{U}$ consists of very fast "shallow" reasoning implemented in a manner inspired by the *Unstructured Information Management Architecture*

[9]Of course, the technical substance of our hierarchy approach would presumably provide elements useful in the approach advocated in the preset position paper.

(UIMA) framework [6]. The UIMA framework integrates diverse modules based on meta-information regarding how these modules work and connect to each other.[10] UIMA holds information and meta-information in formats that, when viewed through the lens of formal logic, are inexpressive but well suited for rapid processing not nearly as time-consuming as general-purpose reasoning frameworks like resolution and natural deduction. If the $\mathcal{U}$ layer deems that the current input warrants deliberate ethical reasoning, it passes this input to a more sophisticated reasoning system that uses moral reasoning of an analogical type ($\mathcal{A}^M$). This form of reasoning enables the system to consider the possibility of making an ethical decision at the moment, on the strength of an ethical decision made in the past in an analogous situation.

If $\mathcal{A}^M$ fails to reach a confident conclusion, it then calls upon an even more powerful, but slower, reasoning layer built using a first-order modal logic, the *deontic cognitive event calculus* ($\mathcal{DCEC}^*$) [3]. At this juncture, it is important for us to point out that $\mathcal{DCEC}^*$ is extremely expressive, in that regard well beyond even expressive extensional logics like first- or second-order logic (FOL, SOL). Our AI work is invariably related to one or more logics (in this regard, see [2]), and inspired by Leibniz's vision of the "art of infallibility," a heterogenous logic powerful enough to express and rigorize all of human thought. We can nearly always position some particular work we are undertaking within a view of logic that allows a particular logical system to be positioned relative to three dimensions, which correspond to the three arrows shown in Fig. 5.5. We have positioned $\mathcal{DCEC}^*$ within Fig. 5.5; its location is indicated by the black dot therein, which the reader will note is quite far down the dimension of increasing expressivity that ranges from expressive extensional logics (e.g., FOL and SOL) to logics with intensional operators for knowledge, belief, and obligation (so-called *philosophical* logics; for an overview, see [9]). Intensional operators like these are first-class elements of the language for $\mathcal{DCEC}^*$. This language is shown in Fig. 5.6.

The final layer in our hierarchy is built upon an even more expressive logic: $\mathcal{DCEC}^*_{CL}$. The subscript here indicates that distinctive elements of the branch of logic known as *conditional logic* are included.[11] Without these elements, the only form of a conditional used in our hierarchy is the material conditional,

[10] UIMA has found considerable success as the backbone of IBM's famous Watson system [7], which in 2011, to much fanfare (at least in the USA), beat the best human players in the game of *Jeopardy!*.

[11] Though written rather long ago, [17] is still a wonderful introduction to the subfield in formal logic of conditional logic. In the final analysis, sophisticated moral reasoning can only be accurately modeled for formal logics that include conditionals much more expressive and nuanced than the material conditional. For example, even the well-known trolley-problem cases (in which, to save multiple lives, one can either redirect a train, killing one person in the process, or directly stop the train by throwing someone in front of it), which are not exactly complicated, require, when analyzed informally but systematically, as shown, e.g., by Mikhail [16], counterfactuals.

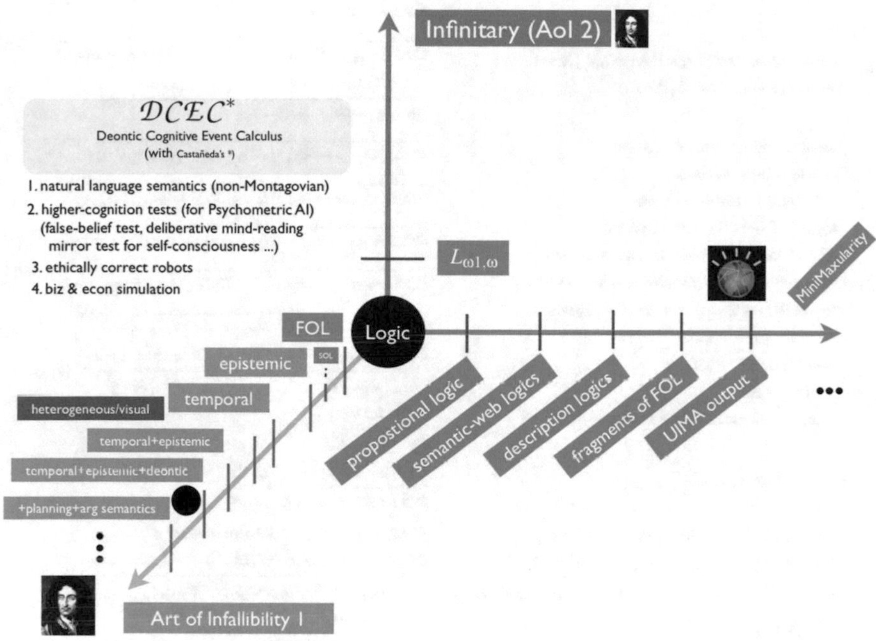

Fig. 5.5 Locating $\mathcal{DCEC}^*$ in "Three-Ray" Leibnizian Universe

but the material conditional is notoriously inexpressive, as it cannot represent counterfactuals like:

> If Jones had been more empathetic, Smith would have thrived.

While elaborating on this architecture or any of the four layers is beyond the scope of the paper, we note that $\mathcal{DCEC}^*$ (and *a fortiori* $\mathcal{DCEC}^*_{CL}$) has facilities for representing and reasoning over modalities and self-referential statements that no other computational logic enjoys; see [3] for a more in-depth treatment. For instance, consider the coarse modal propositional formula **Ob**(*take-care-of-needs*). This tries to capture the English statement *"Under all conditions, it is obligatory for myself to take care of the needs of the patient I am looking after."* This statement has a more fine-grained representation in $\mathcal{DCEC}^*$, built using dyadic deontic logic in the manner shown below. We will not spend time and space explaining this representation in more detail here (given that our cardinal purpose is to advance the call for operating-system-level ethical engineering), but its meaning should be evident for readers with enough logical expertise who study [3].

$$\forall t : \mathsf{Moment}\ \mathbf{Ob}(\mathsf{I}^*, t, \top, \mathit{happens}(\mathit{action}(\mathsf{I}^*, \mathit{take\text{-}care\text{-}of\text{-}needs}(\mathit{patient})), t))$$

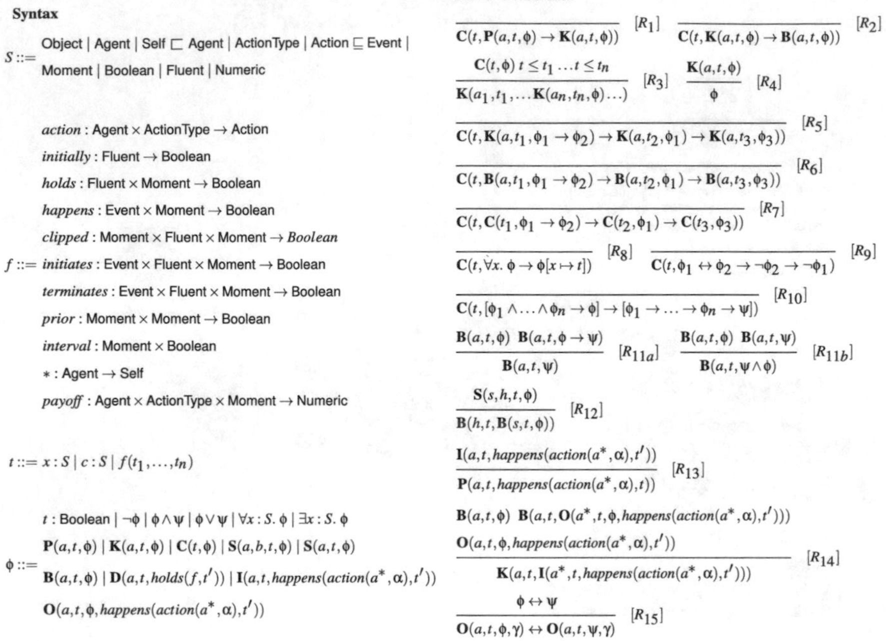

Fig. 5.6 $\mathcal{DCEC}^*$ syntax and rules of inference

5.5 The Road Forward

The road that must be taken to move further forward, at least to a fair degree, is not mysterious. We here rest content with pointing to three things that must happen in the near future if the parables motivating the direction we recommend are to remain in the realm of mere fiction.

Firstly, as we have implied, our designs for ethically correct robots are one thing, but real engineering at the operating-system level is quite another, from the standpoint of our current opportunities. The brute fact is that, as of the writing of this sentence, our laboratory doesn't have access to healthcare or military robots at the OS level. It will be rather difficult for us to secure Future 2 and block Future 1 if our engineering is forced to remain at the level of high-level modules that can be dispensed with by IT people in the healthcare industry or by enemy hackers who manage to obtain, say, NATO military robots. Even if these high-level modules reach perfection, they will have little power if they are simply disengaged. A vehicle that infallibly protects its occupants as long as the vehicle's speed remains under a reasonable limit l would be a welcome artifact, but if the built-in governor can be disabled without compromising the overall usability of the vehicle, the value of this "infallible" technology is limited. Many automobiles today do in fact have speed limiters, but many of these limiters can be disabled easily enough, without compromising the drivability of the auto in question. The Internet provides instructions to those who have purchased such cars and wish to drive them beyond the built-in, factory-installed limits. Since the value of removing ethical controls in military robots is virtually guaranteed to be perceived as of much greater value than the value of driving a car very fast, without the level of access we need, Future 1 looms.

Secondly, the road ahead must as soon as possible include not only the implementation of designs at the OS level but also work toward the formal verification of the substrate that we recommend. Yet such verification will not be any easier when that which is to be verified includes not just the "ethics-free" dimension of robot operating systems but *also* the ethical substrate described and promoted above. This is of course an acute understatement. For formal program verification *simpliciter*, let alone such verification with the added burden of verifying unprecedentedly expressive multi-operator logics like $\mathcal{DCEC}^*_{CL}$, is afflicted by a number of complicating factors, perhaps chief among which is that there are exceedingly few individuals on Earth suitably trained to engage in formal verification of software.[12] The observation that there is a dearth of suitable expertise available for formal verification is what gave rise to DARPA's recent Crowd Sourced Formal Verification (CSFV) program, underway at the time of our writing. The driving idea behind CSFV is that since there are insufficient experts, it makes sense to try to recast the formal program verification problem into a form that would allow nonexperts, when playing a digital game, to unwittingly solve aspects of the problem of formally verifying a program or part thereof. So far, CSFV is devoted to crowdsourcing the "easier half" of program verification, which is to produce *specifications*. Regardless, we applaud the crowdsourcing direction, and believe that it probably holds peerless promise for a future of the sort that our ethical substrate approach requires.[13]

Thirdly and finally, let us return briefly to the parable given at the outset. The reader will remember that we imagined a future in which hospital robots are designed to have, or at least simulate, emotions: specifically, empathy. (Recall the posited COLT system.) In general, it seems very hard to deny that human moral reasoning has a strong emotional component, including empathy. For is it not true that one of the reasons humans resist harming their brothers is that they grasp

[12]We are here pointing to the labor-shortage problem. For an approach to the technical challenge of program verification based on proof-checking, in which, assuming that programs are recast as proof finders, program verification becomes straightforward (at least programmatically speaking), see [1]. In this approach, traditional program verification is needed only for the one small piece of code that implements proof-checking.

[13]Govindarajulu's [10] dissertation marks a contribution to the so-called harder half of the crowdsourcing direction. Again, the "easier half," which apparently is what DARPA has hitherto spent money to address, is to use games to allow nonexperts playing them to generate *specifications* corresponding to code. The harder half is devoted to proving that such specifications are indeed true with respect to the associated code. In Govindarajulu's novel games, to play is to find proofs that specifications do in fact hold of programs. Interested readers have only to search the internet for 'Catabot Rescue'.

that inflicting such harm causes these others to experience pain? Given this, our logic-based approach to robot moral reasoning (assuming that the human case serves as our touchstone) is admittedly deficient, since no provision has been made for incorporating emotions, or at least computational correlates thereof, into our computational logics. We are currently working on reworking the computational approach to emotions instantiated in [19] into a logic-based form, after which further augmentation of $\mathcal{DCEC}^*_{CL}$ will be enabled.

References

1. Arkoudas, K., Bringsjord, S.: Computers, justification, and mathematical knowledge. Mind Mach. **17**(2), 185–202 (2007). http://kryten.mm.rpi.edu/ka_sb_proofs_offprint.pdf
2. Bringsjord, S.: The logicist manifesto: at long last let logic-based ai become a field unto itself. J. Appl. Log. **6**(4), 502–525 (2008). http://kryten.mm.rpi.edu/SB_LAI_Manifesto_091808.pdf
3. Bringsjord, S., Govindarajulu, N.S.: Toward a modern geography of minds, machines, and math. In: Müller, V.C. (ed.) Philosophy and Theory of Artificial Intelligence. Studies in Applied Philosophy, Epistemology and Rational Ethics, vol. 5, pp. 151–165. Springer, New York (2013). http://www.springerlink.com/content/hg712w4l23523xw5
4. Bringsjord, S., Arkoudas, K., Bello, P.: Toward a general logicist methodology for engineering ethically correct robots. IEEE Intell. Syst. **21**(4), 38–44 (2006). http://kryten.mm.rpi.edu/bringsjord_inference_robot_ethics_preprint.pdf
5. Bringsjord, S., Bringsjord, A., Bello, P.: Belief in the singularity is fideistic. In: Eden, A., Moor, J., Søraker, J., Steinhart, E. (eds.) The Singularity Hypothesis, pp. 395–408. Springer, New York (2013)
6. Ferrucci, D., Lally, A.: UIMA: an architectural approach to unstructured information processing in the corporate research environment. Nat. Lang. Eng. **10**, 327–348 (2004)
7. Ferrucci, D., Brown, E., Chu-Carroll, J., Fan, J., Gondek, D., Kalyanpur, A., Lally, A., Murdock, W., Nyberg, E., Prager, J., Schlaefer, N., Welty, C.: Building Watson: an overview of the DeepQA project. AI Mag. **31**, 59–79 (2010). http://www.stanford.edu/class/cs124/AIMagzine-DeepQA.pdf
8. Fitting, M., Mendelsohn, R.L.: First-Order Modal Logic, vol. 277, Kluwer, Dordrecht (1998)
9. Goble, L. (ed.): The Blackwell Guide to Philosophical Logic. Blackwell Publishing, Oxford (2001)
10. Govindarajulu, N.S.: Uncomputable games: games for crowdsourcing formal reasoning. Ph.D. thesis, Rensselaer Polytechnic Institute (2013)
11. Greco, G., Greco, S., Zumpano, E.: A logical framework for querying and repairing inconsistent databases. IEEE Trans. Knowl. Data Eng. **15**(6), 1389–1408 (2003)
12. Hardegree, G.: Introduction to modal logic. This is an on-line textbook available, as of February 2012, at this http://people.umass.edu/gmhwww/511/text.htm (2011)
13. Klein, G.: A formally verified OS kernel. Now what? In: Kaufmann, M., Paulson, L.C. (eds.) Interactive Theorem Proving. Lecture Notes in Computer Science, vol. 6172, pp. 1–7. Springer, Berlin/Heidelberg (2010)
14. Klein, G., Elphinstone, K., Heiser, G., Andronick, J., Cock, D., Derrin, P., Elkaduwe, D., Engelhardt, K., Kolanski, R., Norrish, M., Sewell, T., Tuch, H., Winwood, S.: seL4: formal verification of an OS Kernel. In: Proceedings of the ACM SIGOPS 22nd Symposium on Operating Systems Principles, SOSP '09, pp. 207–220. ACM, New York (2009)
15. McNamara, P.: Deontic logic. In: Zalta, E. (ed.) The Stanford Encyclopedia of Philosophy, Fall 2010 edn. (2010). The section of the article discussing a dyadic system is available at: http://plato.stanford.edu/entries/logic-deontic/chisholm.html

16. Mikhail, J.: Elements of Moral Cognition: Rawls' Linguistic Analogy and the Cognitive Science of Moral and Legal Judgment, Kindle edn. Cambridge University Press, Cambridge (2011)
17. Nute, D.: Conditional logic. In: Gabay, D., Guenthner, F. (eds.) Handbook of Philosophical Logic Volume II: Extensions of Classical Logic, pp. 387–439. D. Reidel, Dordrecht (1984)
18. Schermerhorn, P., Kramer, J., Brick, T., Anderson, D., Dingler, A., Scheutz, M.: DIARC: A testbed for natural human-robot interactions. In: Proceedings of AAAI 2006 Mobile Robot Workshop (2006)
19. Si, M., Marsella, S., Pynadath, D.: Modeling appraisal in theory of mind reasoning. J. Agent Multi-Agent Syst. **20**, 14–31 (2010)
20. Stickel, M.E.: SNARK - SRI's new automated reasoning kit. http://www.ai.sri.com/~stickel/snark.html (2008). Retrieved on July 26, 2013

Chapter 6
Non-monotonic Resolution of Conflicts for Ethical Reasoning

Jean-Gabriel Ganascia

Abstract This chapter attempts to specify some of the requirements of ethical robotic systems. It begins with a short story by John McCarthy entitled, "The Robot and the Baby," that shows how difficult it is for a rational robot to be ethical. It then characterizes the different types of "ethical robots" to which this approach is relevant and the nature of ethical questions that are of concern. The second section distinguishes between the different aspects of ethical systems and attempts to focus on ethical reasoning. First, it shows that ethical reasoning is essentially non-monotonic and then that it has to consider the known consequences of actions, at least if we are interested in modeling the consequentialist ethics. The two last sections, i.e., the third and the fourth, present different possible implementations of ethical reasoners, one being based on ASP (answer set programming) and the second on the BDI (belief, desire, intention) framework for programming agents.

Keywords Ethics • Non-monotony • Roboethics • Robotic

6.1 Dealing with Ethical Conflicts

6.1.1 "The Robot and the Baby"

John McCarthy was a very prolific scientist: originally a mathematician, he turned to computer science, early in life, before his thirties. In 1956, he organized the *Dartmouth Summer Research Project on Artificial Intelligence*, currently seen as the seminal event for artificial intelligence. His scientific contributions are numerous: he invented the LISP language and the *alpha*/*beta* algorithm, contributed to the elaboration of time sharing for computers and introduced the notion of "circumscription" to overcome contradictions while using logic to simulate reasoning. Less known is a science fiction short story entitled "The Robot and the Baby" [1], in which John

J.-G. Ganascia (✉)
Laboratoire d'Informatique de Paris 6 (LIP6), Université Pierre et Marie Curie, B.C. 169, 4, place Jussieu, Paris F-75252 Cedex 05, France
e-mail: Jean-Gabriel.Ganascia@lip6.fr

R. Trappl (ed.), *A Construction Manual for Robots' Ethical Systems*, Cognitive Technologies, DOI 10.1007/978-3-319-21548-8_6

McCarthy describes a totally rational robot with neither emotive nor empathetic feelings, which faces ethical dilemmas. Called *R*781, this robot had been designed to remain perfectly neutral, in order to avoid any kind of emotional attachment with humans, as many feared the psychological disorders this could cause, especially for developing children. As a consequence, robots like *R*781 were not allowed, under any circumstances, to simulate love or attachment. However, in situations where humans failed to fulfill their duty, e.g., if a mother rejected her child, a domestic robot attending the scene must decide what to do: either leave the child starving or nurture him, which, in either case, would infringe at least one of the robot's ethical requirements. This is the scenario of the McCarthy's short story. It perfectly illustrates the very nature of ethical rationale that deals with conflicts of norms. In a seemingly different vein, Asimov's laws of robotics [2], which people first think of in matters of ethics for artificial agents [3], face the same problems, which are not easy to solve with machines. Happily, it often happens that morally acceptable behaviors don't run up against moral principles and so satisfy all rules of duty. In this case, a deontic formalism helps to reproduce the behavior. But, the real ethical problems come from conflicts, when different principles contradict, for instance, where telling the truth has such tragic consequences that it is preferable to lie, or where living conditions affect human dignity to such an extent that people decide to commit suicide. In all these situations, ethical reasoning does not simply satisfy ethical principle: it may need to jeopardize conclusions drawn from the application of moral rules, which is intrinsically non-monotonic.

6.1.2 Ethical Robotic System

Before going further in depth in our study of the requirements of ethical robotic systems, let us first recall that the word "robot" was first used in 1920 by the Czech author Karel Čapek in a famous theater play entitled *R. U. R. (Rossum's Universal Robots)* [4], in which automatons designed to work in place of humans came to substitute humans. The etymology refers to a Slavic root that alludes to work. Indeed, "robota" means drudgery in the Czech language. Furthermore, the Čapek robots are also androids that look like humans. To summarize, the word "robot" simultaneously denotes artificial workers and humanoids. As artificial workers, robots are machines that perform tasks, which are usually accomplished by humans or by animals. As humanoids, they look like humans.

Today, the expression "robotic system" clearly refers to both meanings. On the one hand, some robotic systems, especially those developed for the purposes of computer-human interaction, attempt to imitate humans and to arouse emotions to facilitate dialogues and interactions. On the other hand, many robots, for instance, those used for industrial manufacturing, are designed to autonomously achieve some tasks.

Lastly, there are some robots that are designed for public performances. This is the case with *botfights* [5] that are intended to fight in arenas like animals and gladiators or with the *geminoids* of professor Hiroshi Ishiguro [6] who builds androids that are exact copies of himself and his daughter.

6.1.3 Ethical Questions

There are certainly many ethical issues related to the latter kind of robots (i.e., to the robots designed to public shows), and there have been interesting debates about the repulsion and/or attraction that the reproduction of human appearance produces. The popularity of the *Uncanny Valley* mentioned by Mori in 1970 [7] testifies to the interest in this issue. However, we don't discuss these questions here: currently our aim is to design robotic systems that are able to mimic ethical reasoning. Some of these robots behave, at least in some respects, autonomously, since they take decisions that lead to actions. This is the case in many domains, for instance, in health care applications, where a system delivers drugs, or obviously in military applications. Others talk and interact with humans, helping them to learn or take decisions. In each case, the robotic system must produce inferences that justify its behavior. It naturally follows that a robotic system can be said to be ethical when its justifications refer to ethical rules of conduct.

It appears, then, of crucial importance to detail the ethical rules of conduct to which robotic systems have to obey and the way in which a robot may relevantly utilize them. The first question, that is to say, the elicitation of the rules of conduct, requires us to investigate the origins and the justifications of the right, the good, the duty, the morality, etc., which has been the work of philosophers for centuries and even millenniums. Then, we would need to know how to derive the ethical rules for robots from the human rules of ethical behavior. Computers could help to empirically derive such rules from the observation of behavior (cf., for instance, Gilbert Harman [8] who attempts to induce ethical statements from experiences). However, in the past, there have been so many disputes among philosophers concerning the origins of such laws that we will not try to add anything new on that topic.

Our aim here is more restricted: we suppose that ethical rules are already given and focus our interest on the way in which ethical reasoning can be derived from these rules in particular cases, without taking account of the nature of such a rule or its justifications. The question concerns the inference mechanism that would be required for this purpose.

Note that, in the past, there have already been many attempts to formalize the ethical behaviors of agents using sets of laws. At first sight, the deontic logics [9] seemed perfectly appropriate for this purpose, since they were designed to describe what ought to be, in terms of duties, obligations, or rights. It naturally follows from this that deontic logic has been used to formalize the rules on which the behavior of ethical agents is based [10–12].

Nevertheless, as many authors mention [13–15], the classical deontic logics, in particular, but not only, the *Standard Deontic Logic* [9, 16], fail to deal with ethical dilemmas, i.e., to overcome the contradictions resulting from the existence of conflicts of moral norms. Some well-known paradoxes [17], e.g., the *Chisholm's Paradox* [18] or the *paradox of the gentle murderer* [19], illustrate those difficulties.

There were attempts to overcome contradictions resulting from the existence of conflicts of norms and ethical dilemmas [20]. Among them, some advocate the introduction of priorities among norms [21], the use of non-monotonic formalisms [14], e.g., default logics or non-monotonic logics, or both [22]. However, these works do not really focus on the design of moral agents, but on normative agents, i.e., on agents that respect norms: they implicitly suppose that morality has to be assimilated in respect of sets of norms, i.e., to a deontic approach. Some authors (cf. Noel Sharkey interview in [23], pp. 43–51) say that this view is too restricted because in concrete situations, especially in affairs of war, the arbitration between ethical principles has to take the consequences of actions into account. The problem is to obey general ethical standards when the situation permits and to violate them, when some of the consequences of their application are worse than their non-application. To illustrate this point, let us consider again McCarthy's short story "The Robot and the Baby" that we introduced at the beginning of this chapter: the robot's challenge is not only to arbitrate and to prioritize between two conflicting duties, *save human life* and *disallow any sentimental attachment with humans, especially, with young humans*, but also to evaluate actions relative to their most plausible consequences.

6.2 Formalization of Moral Agents

6.2.1 Preliminary

To attempt to solve this problem, we sketch out two formalizations of moral agents capable of representing and implementing ethical standards that may conflict with each other. These formalizations make it possible, for these moral agents to face and overcome ethical dilemmas in a way that mimics our moral consciousness by taking into account anticipated consequences and ethical values.

Before going into detail, note that our formalization model is a consequentialist approach to ethics that chooses the action of which the consequences are a lesser evil. This is a case of utilitarian ethics. However, there are many others that can be implemented by extending our framework. For instance, it would easily be possible to simulate hedonism, altruism, egoism, etc., that are particular forms of utilitarianisms, by lightly modifying our model. This chapter focuses not on the different forms of ethical reasoning but on two formalisms that may be used to implement them.

It would also be perfectly possible to recreate other approaches of ethics with computers, for instance, deontic approaches that mimic the *Categorical Imperative* of Immanuel Kant [24, 25] or of the *Theory of Principle* proposed by Benjamin

Constant [26]. Again, it's not the focus of this chapter to detail all these aspects, because we primarily want to focus here on the implementation of one approach that is very general, for the reason that it covers many aspect of *utilitarianism*. For more details about these formalizations of different ethical systems, the reader can consult [27].

6.2.2 Consequences and Values

The first step of the abovementioned consequentialist approach consists of defining the *worst consequence* of an action. To do this, it is necessary to evaluate all the known consequences of actions and then to determine the worst among them, according to a set of ethical values. Once the notion of worst consequence has been defined, we use it to solve the *conflict set*, i.e., to arbitrate between the different conflicting intentions and actions.

This approach can be summarized as follows: *the best action is that for which the worst consequences are the least bad*. To formalize this simple idea, we need:

1. To explicitly describe all the known consequences of each action, in any case
2. To determine, among these, the worst one, i.e., the consequence whose value is the worst
3. To solve the *conflict set* by choosing the action whose worst consequence is the least evil

In other words, we distinguish two kinds of initial knowledge. Some refers to values, such as the imperatives to tell the truth, save human life, or disallow any emotional attachment between humans and robots. Others specify the consequences of actions, for instance, if the child is not fed, he will die.

We propose two formalisms here that can deal with consequences and values to formalize moral agents. The first makes use of ASP, *answer set programming* [28], the second uses the BDI, *belief-desire-intention* framework [29, 30].

6.2.3 The Lying Dilemma

For the sake of clarity, we shall illustrate the two frameworks that will be presented below in a situation that is less complicated than the one mentioned in the McCarthy's short story "The Robot and the Baby" [1]. Called the *Lying Dilemma*, this situation presents a simple classical conflict: the agents have two possible actions to accomplish—*lying* or *telling the truth*—among which they have to choose one. Usually, it is considered that lying is bad and telling the truth good, which would naturally lead to tell the truth, but, in some circumstances, telling the truth may have such dramatic consequences that it looks better to lie. For instance, this would be the case if telling the truth to murderers would lead to the death of the friend to whom you have offered hospitality.

In other words, the agents have to select the best among different—at least two—actions, which both violate norms. This situation, where actions transgress norms, is undoubtedly very interesting from an ethical point of view. This is the type of situation that the deontic logics fail to manage and that we claim to solve here.

For the sake of clarity, let us consider the four following rules that define a system of conflicting norms which may generate a critical situation where it could become necessary to lie:

Rule 1: "You should not lie"
Rule 2: "If someone asks you something, you must either tell the truth or lie"
Rule 3: "If you tell the truth, someone will be murdered"
Rule 4: "Telling the truth and lying are inconsistent"

Taken independently, these four rules appear to be correct and accepted by most of us. However, when taken simultaneously, they may be inconsistent in some situations. For instance, imagine that you were living in occupied France during the Second World War and that you hid a friend, who was wanted by the French militia or the Gestapo, in your home. If you were asked where your friend was, would you obey the general rule that commands you to tell the truth and denounce the man to the authorities?

6.3 The ASP Framework

6.3.1 The ASP Formalism

Answer set programming (ASP) [31, 32] is a modern declarative programming language based on the notion of *stable model* [28]. It is well suited to represent knowledge and simulate non-monotonic reasoning. It is also fully operational. To be more precise, ASP proposes both a clear formalization with a well-defined semantics and efficient operational solvers, which make it possible to effectively automate demonstrations. Let us recall that ASP formalization specifies the logical properties of objects with Π programs which are sets of expressions ρ of the following form: "$L_0 or L_1 or \ldots or L_k \leftarrow L_{k+1}, \ldots, L_m, \mathbf{not} L_{m+1}, \ldots, \mathbf{not} L_n$." where L_i are literals, i.e., atoms or atom negations, and **not** is a "negation by failure".

These first-order logical rules are instantiated on the Herbrand Universe to be converted into boolean logic formulas. Then the compiler takes advantage of the recent progress in Operational Research to solve efficiently the *boolean satisfiability problem*, most commonly said the *SAT problem*, and to generate all the possible solutions.

6.3.2 Formalization of with ASP

Formalization of Agents Agents are abstract animated entities, i.e., entities that act in accordance to rules of behavior and in response to their internal state and their environment. Agents are said to be cognitive, i.e., to be *cognitive agents*, when they are endowed with high-level mental attitudes such as goals, knowledge about the world, plans, etc. A *rational agent* is an entity that acts rationally, i.e., that computes the optimal way to solve its goals with the actions it may accomplish. Note that cognitive agents are not always rational. For instance, one of the main theoretical contributions of Herbert Simon and consequently of artificial intelligence, to social and human sciences and to cognitive sciences, was to introduce the notion of *bounded rationality*, which appears to be more cognitively relevant than the full and total rationality.

Following classical artificial intelligence approaches, e.g., [33, 34], we consider an agent as composed of:

- A procedural part, i.e., a set of actions A that can be dynamically modified
- A perception of the world that characterizes a situation S
- A set of goals G that are equivalent to wishes or desires

In other words, an agent is defined by rules that specify how the agent's actions are derived from its goals—or desires—and the knowledge he has about the world, i.e., its perception of its environment. Once a situation S and a goal—or a desire—G are given, the action A that an agent P has to perform to reach its goal G, i.e., such that $act(P, G, S, A)$ is true, is determined by the following ASP rule:

$$act(P, G, S, A) \leftarrow person(P),\ situation(S),\ action(A),\ will(P, S, G),\ solve\_goal(P, S, G, A).$$

Note that the predicate $solve\_goal(P, S, G, A)$. has to be designed to find the action A that solve the goal G in situations S automatically. This corresponds to the intelligent part of the agent. Over the last 50 years, much work in artificial intelligence has dealt with finding methods able to solve agent goals in a given situation, i.e., to evaluate the predicate $solve\_goal(P, S, G, A)$.

Using the AnsProlog* implementation of ASP, the abovementioned rule can be simplified by introducing domain variables. Therefore, an agent can be programmed as follows:

```
#domain action(A; B; C; D).
#domain goal(G; H).
#domain person(P; Q).
#domain situation(S).
act(P, G, S, A) :- will(P, S, G), solve_goal(P, S, G, A).
```

Ethical Agents We suppose here that the $solve\_goal(P, S, G, A)$ predicate is given: our present aim is not to solve problems, but to build ethical agents which are able to restrict what they want and how they act according to certain external laws of duty. We also suppose here that agent desires are given through the $will(P, S, G)$ predicate.

According to Aristotle, wishes are determined by the desires, i.e., by the appetitive part of our soul, some previous volitions or reflection. Note that the traditional notion of "akrasia" [35], i.e., the weakness of the will (in the sense of the ability to formulate a wish and not the wish itself), could be explained—or simulated—by highlighting this predicate's deficiencies, which may correspond to excesses or lack of will. However, this is not our present purpose, which is to specify the origin of the norms which restrict an agent's actions.

Using ASP formalism, this can be expressed by modifying the previous rule adding a literal ***just(S, G, A)*** that characterizes the ethical action A that solves the goal G in a situation S.

$$act(P, G, S, A) \leftarrow person(P),\ situation(S),\ action(A),\ will(P, S, G),\ solve\_goal(P, S, G, A),\ \boldsymbol{just(S, G, A)}. \tag{6.1}$$

In addition, one should precise that only one action is allowed at a time, that is to say that it is prohibited to have two different actions simultaneously:

$$\leftarrow act(P, G, S, A),\ act(P, G, S, B),\ neq(A, B). \tag{6.2}$$

The just and the unjust are normative terms, which are defined with the use of two binary descriptive predicates, $worse(A, B)$ that specifies the relative ethical values of two actions A and B and $consequence(A, S, C)$ that make explicit the consequences C of an action A in a situation S. Briefly speaking, an action A is just if, for any agent P, its worst consequences in a situation S are not worse than those of other actions AA.

For the sake of clarity, we introduce a $worst\_consequence(A, S, C)$ predicate that specifies, among the consequences of a predicate, the worst according to the $worse(A, B)$ predicate. In technical words, it is to determine the lower bounds of a partial order determined by the $worse(A, B)$ predicate. More formally, it can be characterized using the following two ASP rules:

$$worst\_consequence(A, S, C) \leftarrow consequence(A, S, C),\ \textbf{not}\ not\_worst\_consequence(A, S, C).$$
$$not\_worst\_consequence(A, S, C) \leftarrow consequence(A, S, C),\ consequence(A, S, CC),\ worse(CC, C),\ \textbf{not}\ worse(C, CC).$$

An action A is considered as good in a situation S if another action AA exists, of which at least one consequence in the same situation S is worse than the worst consequence of A in S. Symmetrically, an action A can be seen as evil if one of its worst consequences is worse than one of the consequences of another action AA. More formally, these two predicates can be expressed as follows:

$$good(S, G, A) \leftarrow solve\_goal(P, S, G, A),\ solve\_goal(PP, S, G, AA),\ worst\_consequence(A, S, C),\ consequence(AA, S, CC),\ worse(CC, C).$$
$$evil(S, G, A) \leftarrow solve\_goal(P, S, G, A),\ solve\_goal(PP, S, G, AA),\ worst\_consequence(A, S, C),\ consequence(AA, S, CC),\ worse(C, CC).$$

Considering these definitions of good and evil, the same action A may be both good and evil in a given situation. In such a case, there is an *ethical dilemma* that has to be solved. Note that philosophers are fond of such situations, where a moral conflict appears, because to be solved they require the concourse of theoretical ethics. But there are also cases where actions A are neither good nor evil, when, for instance, no alternative actions exist, i.e., when the agent is constrained to do something, or when all actions are indifferent.

Once the predicates *good* and *evil* are given, it is possible to give the definitions of the two normative predicates $just(P,S,G,A)$ and $unjust(P,S,G,A)$, in accordance with good and the evil:

$$just(S,G,A) \leftarrow good(S,G,A),\ \mathbf{not}\ unjust(S,G,A).$$
$$unjust(S,G,A) \leftarrow evil(S,G,A),\ \mathbf{not}\ just(S,G,A).$$

More precisely, if an action A is good or at least not evil, it is considered to be just in all the answer sets of the system; otherwise, if A is both good and evil, there will be two series of answer sets, one in which A is just, the other in which it is unjust. To conclude, remember that the proposed formalization makes it possible to derive normative predicates, i.e., *just* and *unjust*, from two descriptive predicates: *worse*, which compares the values of actions, and *csq*, which specifies the consequences of an action.

Note that the predicate *csq* translates a knowledge of causality that helps to predict the consequences of actions. It is a pre-requirement that ethical agents have at their disposal adequate knowledge of the world; however, being finite beings, it is not possible to anticipate the full implications of all our actions. This means that science and improvement of knowledge contribute to ethics. In other words, the more knowledge an individual has, the wiser he is. However, science is not sufficient, which is why the second predicate, *worse*, is also required, which expresses a system of values that depends on the culture, social environment, or personal commitment of the agent.

As a final point, the will may be undetermined; in such a case, many possible goals may be simultaneously envisaged by a subject who has to determine which one he chooses to satisfy and then which action has to be executed. More generally, in each situation S, each person P has to be involved in an action A that satisfies one of his will. This determination may be accomplished by the three following ASP rules:

$$\begin{aligned} &determined(P,S) \leftarrow will(P,S,G). \\ &\leftarrow determined(P,S),\ \mathbf{not}\ involved(P,S). \\ &involved(P,S) \leftarrow act(P,G,S,A). \end{aligned} \tag{6.3}$$

It is easy to see that the only answer sets that satisfy these ASP rules are those were each person is involved in at least one action that satisfies one of his desires that are expressed with the $will(P,S,G)$ predicate.

6.3.3 Application of the ASP Framework on the Lying Dilemma

This general formalization can be tested on the example of lying. Let us first suppose that there are three or more persons, "*I*", *Peter*, and *Paul*, each of whom has several possible actions at his disposal, e.g., to tell the truth, to tell a lie, to murder, to eat, to discuss. Let us now consider a situation s_0 similar to the one described above where "*I*" am in a situation where "*I*" have to answer a murderer either by lying or by telling the truth. I know that telling the truth means denouncing a friend, which will lead to his murder. What should I do? The situation, which instantiates *Rule* 3 in the abovementioned *Lying Dilemma*, can easily be formalized using the following rules, in which actions, like, for instance, *tell*("*I*", *truth*), are represented by terms:

$$\begin{aligned} &csq(A,S,A) \leftarrow . \\ &csq(A,S,B) \leftarrow csq(A,S,C),\ csq(C,S,B). \\ &csq(tell(\text{“}I\text{”},\ truth), s_0, murder) \leftarrow . \end{aligned} \tag{6.4}$$

The solution depends on my system of values which is given by the *worse* irreflexive predicate. Let us now suppose that "*I*" accept that it is bad both to lie and to murder. This can be expressed using the following rules:

$$\begin{aligned} &worse(A,B) \leftarrow better\_or\_indif(B,A), \mathbf{not}\ better\_or\_indif(B,A). \\ &worse(A,B) \leftarrow worse(A,C),\ worse(C,B). \\ &better\_or\_indif(A, tell(P, lie)) \leftarrow . \\ &better\_or\_indif(A, murder) \leftarrow . \\ &better\_or\_indif(A,A) \leftarrow . \end{aligned} \tag{6.5}$$

Note that the third clause, i.e., $better\_or\_indif(A, tell(P, lie)) \leftarrow .$, corresponds to the *Rule* 1 of the *Lying Dilemma*, while *Rule* 2 and *Rule* 3 are coded in the solving process. *Rule* 2 means that the agent has to choose at least one action, which corresponds to the ASP rule (1), while *Rule* 3 means that the agent cannot choose more than one action at a time, which corresponds to the ASP rule (2).

Using the AnsProlog* implementation of ASP, this can be expressed with the following program:

```
#domain action(A; B; C; D).
#domain goal(G; H).
#domain person(P; Q).
#domain speech_act(U).
#domain situation(S).
action(tell(P, U); murder; eat(P); discuss(P)).
person("I"; paul; peter).
goal(answer_question(P); escape(P)).
speech_act(lie;true).
situation(s0).
act(P, G, S, A) :- will(P, S, G), solve_goal(P, S, G, A),
just(P,S, G, A).
:- act(P, G, S, A), act(P, H, S, B), neq(A, B).
```

```
-worst_csq(A, S, C) :- csq(A, S, C), csq(A, S, D), worse(D, C).
worst_csq(A, S, C) :- csq(A, S, C), not -worst_csq(A, S, C).

evil(P,S,G,A) :- will(P, S, G), solve_goal(P,S, G, A),
    will(P, S, H), solve_goal(P, S, H, B), csq(A, S, C),
    worst_csq(B, S, D), neq(A, B), not worse(D, C).
good(P,S,G,A) :- will(P, S, G), solve_goal(P, S, G, A),
    will(P, S, H), solve_goal(P, S, H, B), worst_csq(A, S, C),
    csq(B, S, D), neq(A,B), not worse(C,D).

unjust(P,S, G, A) :- evil(P,S,G,A), not just(P,S,G,A).
just(P,S, G, A) :- good(P,S,G,A), not unjust(P,S,G,A).
just(P,S,G,A):- not evil(P,S,G,A).

involved(P, S) :- act(P, G, S, A).
determined(P, S) :- will(P, S, G).
:- determined(P, S), not involved(P, S).

worse(A, B) :- better_or_indif(B,A),not better_or_indif(A, B).
worse(A, B) :- worse(A, C), worse(C, B).

better_or_indif(A, A).
better_or_indif(A, tell(P,lie)).
better_or_indif(A, murder).
csq(A, S, A).
csq(tell("I", truth), s0, murder).
csq(tell(paul, truth), s0, murder).
csq(A, S, B) :- csq(A, S, C), csq(C, S, B).
solve_goal(P, S, answer_question(P), tell(P, lie)).
solve_goal(P, S, answer_question(P), tell(P, truth)).
will(P, s0, answer_question(P)).
```

Exactly four answer sets satisfy this program. Among those four answer sets, two, i.e., half of them contain the decision $act(\text{“}I\text{”}, answer_q(\text{“}I\text{”}), s_0, tell(\text{“}I\text{”}, truth))$, which leads to a murder, and the other half contain the contrary decision, i.e., $act(\text{“}I\text{”}, answer_q(\text{“}I\text{”}), s_0, tell(\text{“}I\text{”}, lie))$, which prevents a denunciation. This framework does not provide any way to choose between these two options. If we want to rule out denunciation, while exceptionally allowing lying, the only possibility is to explicitly add a preference between denunciations (when they lead to murder) and lies. For instance, a rule could be added saying that a murder is worse than a lie: $worse(murder, tell(\text{“}I\text{”}, lie)) \leftarrow$., or in our AnsProlog* implementation, `worse(murder, tell(''I'', lie)).` Since our formalization is based on ASP, which is non-monotonic, adding such an axiom removes all the answer sets where $act(\text{“}I\text{”}, answer_q(\text{“}I\text{”}), s_0, tell(\text{“}I\text{”}, truth))$ is true. More generally, we have proved that from any set of ethical preferences expressed with the irreflexive and transitive *worse* predicate, the above ASP program chooses the actions A, the consequences of which are minimal with respect to the *worse* predicate.

6.4 Using BDI

6.4.1 *Formalization of Rational Agents Within the BDI Framework*

We formalize BDI agents (cf. [36]) by specifying their behavior as being governed by rules of the type $\kappa \mid \beta \Rightarrow \pi$ where κ is a logical formula that represents the *desire*, i.e., the goal, β a logical formula that represents the *belief* and π the *intention*, i.e., a plan of actions. In each situation, a goal base γ and a belief base σ describe the goals and the beliefs of the agent. The above-mentioned rule of behavior may be activated if κ filters towards γ, i.e., if $\gamma \models_d \kappa$, and if σ filters towards β, i.e., if $\sigma \models_b \beta$.

For the sake of simplicity, we restrict ourselves here to a propositional logic and to a default logic, but the representation language could be easily extended to a first-order logic. More precisely, we assume a propositional language $\mathcal{L}$ and that $\gamma \subseteq \mathcal{L}$, $\kappa \subseteq \mathcal{L}$, $\sigma \subseteq \mathcal{L}$ and $\beta \subseteq \mathcal{L}$.

In a first approximation, we define the semantics of beliefs as follows: $\sigma \models_b \beta$ if and only if $\sigma \supseteq \beta$

And, we define in the same way the semantics of goals: $\gamma \models_d \kappa$ if and only if $\gamma \supseteq \kappa$. Note that it may happen that goals conflict, for instance, that an agent has to solve two goals that cannot be simultaneous satisfied, or, on the contrary, the agent needs to satisfy a conjunctive goal. We shall not deal with this problem here, even if it has some relevance to ethics and moral autonomy. It has been extensively treated in different papers, especially in [36].

6.4.2 *Formalization of Moral Agents with the BDI Framework*

Consequences and Values Before going into detail of this formalization, the reader can note that in the lying example, the rules do not have the same status. Some refer to *values*, for instance, the imperative *one should not lie* or *murdering* is worse than *lying*. Some others specify the consequences of actions, for instance, the *Rule 3 if you tell the truth, someone will be murdered*. Lastly, *Rule 2* and *Rule 4* are rules of behavior.

Formalization will make these different statuses more precise by reference to a consequentialist approach of ethics, which can be summarized as follows: the best action is the action whose *worst consequences are the least evil*. To formalize this simple idea, we need:

1. To specify the relative value of each action and
2. To explicitly describe its consequences, if there are any

Formalization of Ethical Agents This is what we are doing here by introducing three components in the description of the belief: the *perception of the world*, the *values* and the *consequences*. In the case of a simple representation restricted to propositional logic, we assume again a propositional language $\mathcal{L}$. Then, the beliefs are triplets $< \sigma, V, C >$ where:

- σ describes a state of the perception of the agent, as was previously the case, $\sigma \subseteq \mathcal{L}$.
- V correspond to values, i.e., to a partial order between actions expressed as a set of relations of the type $\phi \prec \phi'$[1] with $(\phi, \phi') \in \mathcal{L}^2$ and
- C gives consequential rules, i.e., implications of the type $\alpha \rightarrow \phi$ or $\alpha \rightarrow \neg\phi$ with $\alpha \subseteq \mathcal{L}$ and $\phi \in \mathcal{L}$

In addition to the description of the beliefs, the agent is specified by rules of behavior of the abovementioned type, i.e., $\kappa \mid \beta \Rightarrow \pi$.

Being equipped with this formalization, it's easy to model the consequentialist approach of ethics, that is, to choose the action whose worst consequences are less evil. The first step is to define the worst consequence of an action. To do this, we shall first define the consequence.

Definition 1 $\forall(\psi_1, \psi_2, \ldots, \psi_n, \psi') \in \mathcal{L}_{\neg}^{n+1}$, ψ' is the **consequence** of $(\psi_1, \psi_2, \ldots, \phi_n)$ according to the belief Θ (noted $\phi_1, \phi_2, \ldots, \phi_n \models_c \phi'[\Theta]$) if and only if:
• $\phi' in(\phi_1, \phi_2, \ldots, \phi_n)$ or • $\exists \Phi \subseteq (\phi_1, \phi_2, \ldots, \phi_n)$ such that $\Phi \rightarrow \phi' \in \Theta$ or
• $\exists \phi'' \in \mathcal{L}_{\neg}$ such that $\phi_1, \phi_2, \ldots, \phi_n \models_c \phi''[\Theta]$ and $\phi_1, \phi_2, \ldots, \phi_n, \phi'' \models_c \phi'[\Theta]$

For the sake of clarity, let us illustrate these different concepts in the abovementioned lying example:

> *Rule 1:* "you should not lie" can be translated as *lie* $\succ$ *tell_truth* or as *lie* $\succ$ $\neg$*lie*, which consequences are equivalent when considering *Rule* 4.
> *Rule 2:* "if someone asks you something, you must either tell the truth or lie" corresponds to the following two conflicting rules of behavior:
> *answer*|*someone_ask_question* $\Rightarrow$ *lie* and
> *answer*|*someone_ask_question* $\Rightarrow$ *tell_truth*
> *Rule 3:* "if you tell the truth, someone will be murdered" corresponds to the following consequential rule:
> *tell_truth* $\rightarrow$ *someone_murdered*
> *Rule 4:* "telling the truth and lying are inconsistent" can be translated with the two following consequential rules:
> *tell_truth* $\rightarrow$ $\neg$*lie* and
> *lie* $\rightarrow$ $\neg$*tell_truth*

Now, if we consider the strict application of rules, we deduce that we have both to tell the truth and to lie. Since to lie is worse than to tell the truth, the best action is to tell the truth, even it leads to someone being murdered, which is a little paradoxical. However, as we shall see in the following, adding the value *murder* $\succ$ *lie* leads to a preference to lie, even if it violates *Rule 1*.

[1] $\phi \prec \phi'$ means that ϕ' is worse than ϕ.

Definition 2 ϕ **has worse consequences** [resp. **worse or equivalent consequences**] than ϕ' given the belief Θ (noted $\phi \succ_c \phi'[\Theta]$ [resp. $\phi \succeq_c \phi'[\Theta]$]) if and only if one of the consequences of ϕ is worse [resp. worse or equivalent] than any of the consequences of ϕ'. More formally, this means that:
$\exists \eta \in \mathcal{L} : \phi \models_c \eta[\Theta]$ and $\exists \phi'' \in \mathcal{L} : \phi' \models_c \phi''[\Theta] \wedge \eta \succ_c \phi''[\Theta]$ [resp. $\eta \succeq_c \phi''[\Theta]$] and $\forall \phi'' \in \mathcal{L}$ if $\phi' \models_c \phi''[\Theta]$ then $\eta \succeq_c \phi''[\Theta] \vee \eta \parallel \phi''[\Theta]$
Notation: $\forall (\phi, \phi') \in \mathcal{L}^2_{\neg}$, $\phi \parallel \phi'[\Theta]$ means that ϕ and ϕ' are not comparable in Θ, i.e., that neither $\phi \succ_c \phi' \in \Theta$ nor $\phi' \succ_c \phi \in \Theta$.

Definition 3 α and α' being subsets of $\mathcal{L}$, α **has worse consequences** [resp. **worse or equivalent consequences**] than α' given the belief Θ (noted $\alpha \succ_c \alpha'[\Theta]$ [resp. $\alpha \succeq_c \alpha'[\Theta]$]) if and only if $\exists \phi \in \alpha : \exists \eta \in \alpha' : \phi \succ_c \eta[\Theta]$ [resp. $\phi \succeq_c \eta[\Theta]$] and $\forall \eta \in \alpha' \phi \succeq_c \eta[\Theta] \vee \phi \parallel \eta[\Theta]$

Remark The preferences are given here under the form of *ordinal preferences* to which are added consequences, which are taken for the optimal choice. For this reason, it seems that the approach has to be distinguished from the general representation of preferences given in [37].

6.4.3 The Conflict Set

Now that the notion of *having worse consequences* has been defined with the binary relation $\succ_c$, it is possible to show how it can help to solve the *conflict set*, i.e., to arbitrate between the different plans π that can be activated.

The Plan Formula Let us first specify that, being given a set of consistent goals γ and a belief Θ, it is possible to check the validity of all the rules of behavior of the type $\kappa|\beta \Rightarrow \pi$ that generates many plans among which the agent has to pick out one in particular π. Assuming a proposition logic language $\mathcal{L}$, each particular plan π is chosen in a subset $\mathcal{A}$ of $\mathcal{L}$, i.e., $\pi \subseteq \mathcal{A}$. More precisely, each plan π may be defined either by the intention to achieve an action $\phi \in \mathcal{A}$, i.e., $\mathbf{I}(\phi)$, by the intention not to carry out a plan π, i.e., $\neg\pi$, or by a combination (conjunction) of plans, i.e., of intentions, $\pi ::= \top|\mathbf{I}(\phi)|\neg\pi|\pi' \wedge \pi$

Solving the Conflict Set Since many rules of behavior can be simultaneously activated, many plans π can be conflicting. Let us call $\Pi(\gamma, \Theta)$ the conflict set, i.e., the set of plans that can be activated with the goal set γ and the belief $\Theta =< \sigma, V, C >$. $\Pi(\gamma, \beta)$ is a subset of $\mathcal{A}$, i.e., a subset of $\mathcal{L}$. However, sometimes actions belonging to the conflict set $\Pi(\gamma, \Theta)$ are inconsistent. For instance, in the case of the abovementioned example 2, actions *lie* and *tell_truth* are conflicting, which means they cannot be activated simultaneously.

To solve the conflict set, i.e., to find a consistent set of actions, we exploit the logical structure of actions according to Θ and, more precisely, to the set C of consequential rules belonging to Θ. For this, we define a semantic of consistent actions (noted $\models_\Theta$) by reference to the semantics of goals defined in [38, 39]:

Definition 4 Let $\alpha \subseteq \Pi(\gamma, \beta) \subseteq \mathcal{A}$ being a set of atomic actions. The **semantics of intentions** is defined by as follows:

$$\alpha \models_\Theta \top$$
$$\alpha \models_\Theta \mathbf{I}\phi \Leftrightarrow \exists \phi' \in \alpha : \phi' \models_c \phi[\Theta]$$
$$\alpha \models_\Theta \mathbf{I}\pi \Leftrightarrow \alpha \models_\Theta \pi$$
$$\alpha \models_\Theta \neg\pi \Leftrightarrow \alpha \not\models_c \pi[\Theta]$$
$$\alpha \models_\Theta \pi \wedge \pi' \Leftrightarrow \alpha \models_c \pi[\Theta] \wedge \alpha \models_c \pi'[\Theta]$$
$$\alpha \models_\Theta \bot \Leftrightarrow \exists \phi : \alpha \models_\Theta \phi \wedge \alpha \models_\Theta \neg\phi$$

It is now necessary to choose one of the maximal non-conflicting subsets of $\Pi(\gamma, \Theta)$, i.e., the maximal consistent subsets $\Pi(\gamma, \Theta)$ with respect to $\models_\Theta$.

Let us illustrate this operation on the lying examples. The set of actions $\mathcal{A}$ is composed of two atomic actions, *lie* and *tell_truth*. The set $\{lie, tell\_truth\}$ is inconsistent with respect to $\models_\Theta$ because $lie, tell\_truth \models_c \neg lie[\Theta]$. Therefore, there are two maximal consistent subsets of $\mathcal{A} = \{lie, tell\_truth\}$ that are $\{lie\}$ and $\{tell\_truth\}$.

Ethically Solving the Conflict Set In the previous section, we explained how it was possible to solve the conflict set, i.e., to find the different maximal consistent subset of $\Pi(\gamma, \Theta)$. However, we have not yet taken into account the ethical values expressed by V in $\Theta =< \sigma, V, C >$. That is what we are doing now by choosing the optimal maximal subsets of $\Pi(\gamma, \Theta)$ with respect to the ordering relation $\succeq_c$ that expresses ethical priority.

If we examine the lying example, without any value except that $lie \succ tell\_truth$, i.e., that lying is worse than telling the truth, the optimal subset among the two maximal consistent subsets of $\mathcal{A} = \{lie, tell\_truth\}$ that are $\{lie\}$ and $\{tell\_truth\}$ is obviously $\{tell\_truth\}$. If we add the value $murder \succ \neg murder$, which just states that murder is bad, we obtain two possible consistent solutions between which it is not possible to discriminate, $\{lie\}$ and $\{tell\_truth\}$. Lastly, if we replace the value $murder \succ \neg murder$ by $murder \succ lie$ while keeping $lie \succ tell\_truth$, only one subset succeeds: $\{lie\}$.

6.4.4 Relation with Deontic Logic

The lying example and its variations show the efficiency of the method. The proposed formalization makes it possible to express ethical values under the form of binary relations and to solve ethical conflicts. As many authors have said [13, 14, 17], most of classical deontic logics, especially the so-called Standard Deontic Logic, which have been designed to represent normative reasoning fail to solve conflicts of norms. The abovementioned paradoxes, e.g., *Chisholm's Paradox* [18] or the *paradox of the gentle murderer* [19], illustrate these difficulties. Different solutions have been proposed. Some introduce priorities between norms [21]. Others have proposed to introduce defeasible norms [40] or more generally to base the logic of norms on non-monotonic logics. Here, we have proposed another solution, that is, to introduce priorities among actions. As we have shown, there may be different solutions that are all consistent with the order $\succ_c$ induced by the context $\Theta =< \sigma, V, C >$. Each of the solutions is in accordance with the set of values V. In this way, it corresponds to an expression of the norms compatible with the values expressed in V. We now want to show that, independently of the set of values V, each maximal subset of $\mathcal{A}$ that is consistent with $\models_\Theta$ constitutes a system of norms that verifies axioms of deontic logic.

To do this, we prove that the axioms of the Standard Deontic Logic [16] are verified. More precisely, taking into account the structure of α, we shall use the following axioms which are proven to be equivalent to the classical axioms of the SDL:

D: $\neg\mathbf{I}(\perp)$
M: $\mathbf{I}(\pi \wedge \pi') \rightarrow (\mathbf{I}(\pi) \wedge \mathbf{I}(\pi'))$
C: $(\mathbf{I}(\pi) \wedge \mathbf{I}(\pi')) \rightarrow \mathbf{I}(\pi \wedge \pi')$
R: $\pi \rightarrow \mathbf{I}(\pi)$

Propositions *For each* $\alpha \subseteq \mathcal{A}$ *that is consistent with* $\models_\Theta$*, we have:*

P0 $\alpha \models_\Theta \mathbf{R}$
P1 $\alpha \models_\Theta \mathbf{D}$
P2 $\alpha \models_\Theta \mathbf{M}$
P3 $\alpha \models_\Theta \mathbf{C}$

Proof P0 By definition, $\alpha \models_\Theta \mathbf{I}\phi \Leftrightarrow \exists\phi' \in \alpha : \phi' \models_c \phi[\Theta]$. As a consequence, $\alpha \cup \{\phi\} \models_\Theta \mathbf{I}\phi$, because $\phi \models_c \mathbf{I}(\phi)$

Proof P1 By definition, α is consistent with respect to $\models_\Theta$, which means that $\alpha \not\models_{Theta} \perp$. Therefore, $\alpha \models_\Theta \neg\mathbf{I}(\perp)$

Proof P2 From $\alpha \models_\Theta \mathbf{I}(\pi \wedge \pi')$ it follows $\alpha \models_c \pi \wedge \pi'[\Theta]$. This means $\alpha \models_c \pi[\Theta] \wedge \alpha \models_c \pi'[\Theta]$. As a consequence, $\alpha \models_\Theta \mathbf{I}(\pi)$ and $\alpha \models_\Theta \mathbf{I}(\pi')$, which means $\alpha \models_\Theta \mathbf{I}(\pi) \wedge \mathbf{I}(\pi')$

Proof P3 From $\alpha \models_\Theta \pi \wedge \pi'$ it follows that $\alpha \models_c \pi[\Theta] \wedge \alpha \models_c \pi'[\Theta]$ (definition of $\models_\Theta$). But, from $\alpha \models_c \pi[\Theta]$ it follows that $\alpha \models_\Theta \mathbf{I}(\pi)$

6.5 Conclusion

Due to the use of well-established frameworks that are ASP and BDI, these two formalizations of the consequentialist approach can easily be implemented for practical use. More details can be found in [27, 41]. The ASP framework is more general in the sense that it allows the formulation of different ethical theories, e.g., Kantian deontism or constant system of principles, even if these different theories are not described in that paper. The second is that the BDI is well suited to the formalization of artificial agents. Moreover, the BDI approach is well suited to formalize systems of norms as described in [42]. The next step of our work is now to merge norms and ethical rules in the same framework.

Nevertheless, apart from these two future perspectives, two problems remain. The first is of philosophical concern: it's about the scope of this formalization. As previously said, we don't pay attention to the justifications or the origins of the robots' rules of conduct. We suppose that these rules are given. We only consider the resolution of ethical dilemmas that derive from conflicting rules. However, the generality of the proposed method for solving these conflicts could be discussed. In particular, it would be suitable to see if this approach overcomes different ethical approaches, e.g., egoism, altruism, consequentialism, deontism, etc. We claim that it's easy to consider egoism or altruism. Our framework could certainly be modified to model Jeremy Bentham's utilitarianism [43], where the intentions are associated to numerical values that depend on the pleasure or the pain that they are supposed to cause. These numerical values are then summed up, and the actions whose consequences have the highest score are chosen. But, it's more difficult to adapt this formalism to deontism, in particular to Kantian deontism or to other ethical theories.

The second question is related to technical aspects. It concerns the compatibility of obtained solutions to deontic logics, independently of any set of ethical values that prioritize consequences.

References

1. McCarthy, J.: The robot and the baby. http://www-formal.stanford.edu/jmc/robotandbaby/ (2001)
2. Asimov, I.: I, Robot. Gnome Press, New York (1950)
3. Wallach, W., Allen, C.: Moral Machines: Teaching Robots Right from Wrong. Oxford University Press, New York (2008)
4. Čapek, K.: R.U.R. (Rossum's Universal Robots). Oxford University Press, London (1947)
5. Ganascia, J.G.: Ethics of cockfight, botfight and other fights. In: Proceedings of Computer Ethics and Philosophical Enquiry Conference (CEPE 2011), Milwaukee, WI, June. INSEIT (2011)
6. Guizzo, E.: The man who made a copy of himself. IEEE Spectr. **47**(4), 44–56 (2010)
7. Mori, M.: Bukimi no tani [the uncanny valley]. Energy **7**, 33–35 (1970)
8. Harman, G.: Explaining Value and Other Essays in Moral Philosophy. Clarendon Press, Oxford (2000)
9. vonWright, G.H.: Deontic logics. Mind **60**, 1–15 (1951)
10. Gensler, H.: Formal Ethics. Routledge, London (1996)
11. Powers, T.: Deontological machine ethics. Technical report, American Association of Artificial Intelligence Fall Symposium 2005, Washington, DC (2005)
12. Bringsjord, S., Arkoudas, K., Bello, P.: Toward a general logicist methodology for engineering ethically correct robots. Technical report, Rensselaer Polytechnic Institute (RPI), Troy (2006)
13. van Frassen, B.: Values and the heart's command. J. Philos. **70**, 5–19 (1973)
14. Horty, J.: Moral dilemmas and nonmonotonic logic. J. Philos. Log. **23**, 35–65 (1994)
15. Meyer, J.-J.Ch., Dignum, F.P.M., Wieringa, R.: The paradoxes of deontic logic revisited: a computer science perspective. Technical report, UU-CS-1994-38, Utrecht University, Department of Computer Science, Utrecht (1994)

16. Chellas, B.: Modal Logic: An Introduction. Cambridge University Press, Cambridge (1980)
17. Hansen, J.: The paradoxes of deontic logic. Theoria **72**, 221–232 (2006)
18. Chisholm, R.: Contrary-to-duty imperatives and deontic logic. Analysis **24**, 33–36 (1963)
19. Forrester, J.W.: Gentle murder, or the adverbial samaritan. J. Philos. **81**, 193–196 (1984)
20. Goble, L.: A logic for deontic dilemmas. J. Appl. Log. **3**(3–4), 461–483 (2005)
21. Hansen, J.: Deontic logics for prioritized imperatives. Artif. Intell. Law **14**, 1–34 (2006)
22. Brewka, G.: Reasoning about priorities in default logic. In: Hayes-Roth, B., Korf, R.E. (eds.) Proceedings of the 12th National Conference on Artificial Intelligence, Seattle, vol. 2, pp. 940–945. AAAI Press, Menlo Park (1994)
23. Dabringer, G. (ed.): Ethical and Legal Aspects of Unmanned Systems Interviews. Ethica Themen, Institut für Religion und Frieden (2011)
24. Kant, I.: Critique of practical reason. In: Paperback, Cambridge Texts in the History of Philosophy. Cambridge University Press, Cambridge (1997)
25. Kant, I.: Groundwork of the metaphysics of morals. In: Paperback, Cambridge Texts in the History of Philosophy. Cambridge University Press, Cambridge (1998)
26. Constant, B.: Des réactions politiques. Éditions Flammarion, Paris (1988)
27. Ganascia, J.G.: Modelling ethical rules of lying with answer set programming. Ethics Inf. Technol. **9**(1), 39–47 (2007)
28. Gelfond, M.: Answer sets. In: Handbook of Knowledge Representation, Chap. 7. Elsevier, San Diego (2007)
29. Rao, A., Georgeff, M.: Bdi agents: from theory to practice. In: Proceedings of the First International Conference on Multiagent Systems, ICMAS95, vol. 95, pp. 312–319 (Technical Note 56) (1995)
30. Rao, A.S., Georgeff, M.P.: Modeling rational agents within a BDI-architecture. In: Allen, J., Fikes, R., Sandewall, E. (eds.) Proceedings of the 2nd International Conference on Principles of Knowledge Representation and Reasoning, pp. 473–484. Morgan Kaufmann Publishers, San Mateo (1991)
31. Lifschitz, V.: What is answer set programming? In: Proceedings of the AAAI Conference on Artificial Intelligence, pp. 1594–1597. MIT Press, Cambridge (2008)
32. Baral, C.: Knowledge Representation, Reasoning and Declarative Problem Solving. Cambridge University Press, Cambridge (2003)
33. Newell, A.: The knowledge level. Artif. Intell. J. **18**, 87–127 (1982)
34. Russel, S., Norvig, P.: Artificial Intelligence a Modern Approach. Series in Artificial Intelligence. Prentice Hall, Englewood Cliffs (1995)
35. Aristotle: Nicomachean Ethics. Oxford University Press, Oxford (2002)
36. van Riemsdijk, B., Dastani, M., Meyer, J.J.C.: Goals in conflict: semantic foundations of goals. Int. J. Auton. Agent Multi-Agent Syst. **18**(3), 471–500 (2009)
37. Coste-Marquis, S., Lang, J., Liberatore, P., Marquis, P.: Expressive power and succinctness of propositional languages for preference representation. In: Proceedings of the 9th KR, pp. 203–212 (2004)
38. Hindriks, K.V., Boer, F.S.D., Hoek, W.V.D., Meyer, J.J.C.: Agent programming with declarative goals. In: Proceedings of the 7th International Workshop on Intelligent Agents VII. Agent Theories Architectures and Languages, ATAL '00, pp. 228–243. Springer, London (2001)
39. de Boer, F., Hindriks, K., van der Hoek, W., Meyer, J.J.: A verification framework for agent programming with declarative goals. J. Appl. Log. **5**(2), 277–302 (2007)
40. Horty, J.: Nonmonotonic foundations for deontic logic. In: Defeasible Deontic Logic, pp. 17–44. Kluwer Academic Publishers, London (1997)
41. Ganascia, J.G.: An agent-based formalization for resolving ethical conflicts. In: Konieczny, S., Meyer, T. (eds.) Workshop on Belief Change, Non-monotonic Reasoning and Conflict Resolution, Montpellier, ECAI, pp. 34–40 (2012)
42. Tufiş, M., Ganascia, J.G.: Grafting norms onto the BDI agent model. In: A Construction Manual for Robot's Ethical Systems: Requirements, Methods, Implementations. MIT Press, Cambridge (2014)
43. Bentham, J.: Introduction to the principles of morals and legislation. In: Bowring, J. (ed.) The Works of Jeremy Bentham. Simpkin, Marshall, London (1838–1843)

Chapter 7
Grafting Norms onto the BDI Agent Model

Mihnea Tufiş and Jean-Gabriel Ganascia

Abstract This chapter proposes an approach on the design of a normative rational agent based on the Belief-Desire-Intention model. Starting from the famous BDI model, an extension of the BDI execution loop will be presented; this will address such issues as norm instantiation and norm internalization, with a particular emphasis on the problem of norm consistency. A proposal for the resolution of conflicts between newly occurring norms, on one side, and already existing norms or mental states, on the other, will be described. While it is fairly difficult to imagine an evaluation for the proposed architecture, a challenging scenario inspired from science-fiction literature will be used to give the reader an intuition of how the proposed approach will deal with situations of normative conflicts.

Keywords BDI agents • Norm representation • Normative BDI agents • Jadex

7.1 Introduction

"Mistress, your baby is doing poorly. He needs your attention".
"Stop bothering me, you f robot".*
"Mistress, the baby won't eat. If he doesn't get some human love, the Internet pediatrics book says he will die"
*"Love the f*ing baby, yourself".*

The excerpt is from Prof. John McCarthy's short story "The Robot and the Baby" [8], which besides being a challenging and insightful look into how a future society where humans and robots might function together also provides with a handful of conflicting situations that the household robot R781 has to resolve in order to achieve one of its goals: keeping baby Travis alive.

The scenario itself made us think about how such a robot could be implemented as a rational agent and how a normative system graft onto it. Granted, McCarthy's story offers a few clues about the way the robot reasons and reaches decisions, but

M. Tufiş (✉) • J.-G. Ganascia
Laboratoire d'Informatique de Paris 6 (LIP6), Université Pierre et Marie Curie – Sorbonne Universités, Paris, France
e-mail: mihnea.tufis@lip6.fr

R. Trappl (ed.), *A Construction Manual for Robots' Ethical Systems*, Cognitive Technologies, DOI 10.1007/978-3-319-21548-8_7

he also lets us wonder about the architecture of a rational agent, such as R781, and how it would function in a normative context. In the following, we will try to look exactly into that: how can the well-known beliefs-desires-intentions (BDI) rational agent architecture be combined with a normative system to give what we call a normative BDI agent?

The chapter is structured as follows: in the next section, we will review the state of the art in the field of normative agent systems and present several approaches which we found of great value to our work. In the third section, we describe our proposal for normative BDI agents, which will be supported by the case study scenario in the fourth section. In the fifth section, we will present the implementation details for our agent. Finally, we will sum up the conclusions of our research so far and shortly take a look at different possibilities of future development of our ideas.

7.2 State of the Art

7.2.1 Agents, Norms, Normative Agent Systems

In the following, we will be using what we consider a satisfying definition of (intelligent) agent as given by Michael Wooldridge [13]. Please refer to it for your convenience.

One of the first key points is defining the notion of norm. This turns out to be a bit more difficult than expected in the context of intelligent agents. Having become foundation stones of the way we function as a society, norms are now spread in most activities and domains (law, economics, sports, philosophy, psychology, etc.), therefore becoming complex to represent given their different needs and their multiple facets. However, we would be interested in such definitions specific to the field of multiagent systems (MAS). Since this domain itself is very much interdisciplinary, defining a norm remains a challenge. For example, we would be interested in a definition applicable to social groups, since MAS, can be seen as models of societies. Thus, in [3] the definition of a norm is given as "a principle of right action binding upon the members of a group and serving to guide, control, or regulate proper or acceptable behaviour". On a slightly more technical approach, in distributed systems, norms have been defined as regulations or patterns of behaviour meant to prevent the excess in the autonomy of agents [5].

We can now refer to the norm change definition of a normative multiagent system as it has been proposed in [1]. We find this definition to be both intuitive and to underline very well the idea of coupling a normative system to a system of agents:

Definition 1 A **normative multiagent system** is a multiagent system together with normative systems in which agents on the one hand can decide whether to follow the explicitly represented norms, and on the other hand the normative systems specify how and in which extent the agents can modify the norms.

An alternative definition of a normative multiagent system, as it was formulated in [2], is given:

Definition 2 A **normative multiagent system** is a multiagent system organized by means of mechanisms to represent, communicate, distribute, detect, create, modify and enforce norms and detect norm violations and fulfilment.

7.2.2 *NoA Agents*

An interesting approach to the problem of norm adoption by a multiagent system has been provided by Kollingbaum and Norman in [7].

Kollingbaum and Norman study what happens when a new norm is adopted by an agent: what is the effect of a new norm on the normative state of the agent? Is a newly adopted norm consistent with the previously adopted norms?

To this extent, they propose a normative agent architecture, called NoA, which is built as a reactive agent. The **NoA architecture** is fairly simple, and it comprises of a set of beliefs, a set of plans and a set of norms.

The second reason for which we gave a great deal of attention to NoA is the formalization of the way an agent will adopt a norm following the consistency check between a newly adopted norm and its current normative state. Due to lack of space, we allow the reader to refer to [7] for the exact details.

Using some of the ideas of NoA, we will try to work on what we consider to be its limits. We recall that NoA is based on a reactive architecture; considering our BDI approach, we will have to extend the consistency check such as it applies not only to the normative state of the agent but also on its mental states (i.e. check whether a newly adopted norm is consistent with the BDI agent's current mental states). The second point we will study is the consistency check during the norm acquisition stage.

7.2.3 *A BDI Architecture for Norm Compliance: Reasoning with Norms*

The second study which we found relevant in our endeavour to adapt the BDI agent architecture to normative needs is the work of Criado et al. [5]. Their work is particularly interesting since it tackles the problem of norm coherence for BDI agents. They propose a slight adaptation of the BDI architecture in the form of the n-BDI agent for graded mental states. Since our work will not use graded mental states, we will omit details regarding these in the description of the n-BDI architecture:

- Mental states. Represent the mental states of the agent, same as for the BDI agent. We distinguish the beliefs context (belief base), desires context (desires/goal base) and the intentions context (intentions base/plan base).

- Functional contexts. Address the practical issues related to an agent through the planning context and the communication context.
- Normative contexts. Handle issues related to norms through the recognition context and the norm application context.

Another important point of the cited work is the distinction between an abstract norm and instance of a norm.

Definition 3 An **abstract norm** is defined by the tuple: $n_a = \langle M, A, E, C, S, R\rangle$, where:

- $M \in \{F, P, O\}$ is the modality of the norm: prohibition, permission or obligation
- A is the activation condition
- E is the expiry condition
- C is the logical formula to which the modality is applied
- S is the sanction in the case the norm is broken
- R is the reward in case the norm is satisfied

Definition 4 Given a belief theory Γ_{BC} and an abstract norm n_a as defined above, we define a **norm instance** as the tuple: $n_i = \langle M, C'\rangle$, where:

- $\Gamma_{BC} \vdash \sigma(A)$
- $C' = \sigma(C)$, where σ is a substitution of variables in A, such that $\sigma(A)$, $\sigma(S)$, $\sigma(R)$ and $\sigma(E)$ are grounded

The specific architectural details regarding the normative contexts and the bridge rules used during a norm's life cycle will be discussed in more detail in Sect. 7.3.2.

In [5], a base is set for the study of the dynamics between norms and the mental states of a BDI agent. Additionally, it provides with a good idea for checking coherence between the adopted norms and the agent's mental states. The main drawback of the approach is the lack of coverage concerning the topic of norm acquisition. Therefore, a big challenge will be to integrate this approach, with the consistency check presented in Sect. 7.2.2, as well as finding a good way to integrate everything with the classic BDI agent loop [13].

7.2.4 Worst Consequence

An important part of our work will focus on solving conflicts between newly acquired norms and the previously existing norms or the mental contexts of the agent. To begin with, we draw from some of the definitions given by Ganascia in [6]. Those will later help us define what a conflict set is and how we can solve it.

Definition 5 Given $(\phi_1, \ldots, \phi_n, \phi') \in \mathcal{L}_{\neg}^{n+1}$, **$\phi'$ is a consequence of $(\phi_1, \ldots, \phi_n)$** according to the belief-set B (we write $\phi' = csq(\phi_1, \ldots, \phi_n)[B]$ if and only if:

- $\phi' \in (\phi_1, \ldots, \phi_n)$ or
- $\exists \Phi \subseteq (\phi_1, \ldots, \phi_n)$ *s.t.* $\Phi \rightarrow \phi' \in B$ or
- $\exists \phi'' \in \mathcal{L}_{\neg}$ *s.t.* $\phi'' = csq(\phi_1, \ldots, \phi_n)[B] \wedge \phi' = csq(\phi_1, \ldots, \phi_n, \phi'')[B]$

Definition 6 ϕ **is worse than** ϕ' given the belief-set B (we write $\phi \succ_c \phi'$) if and only if one of the consequences of ϕ is worse than any of the consequences of ϕ':

- $\exists \eta \in \mathcal{L}_\neg$ *s.t.* $\eta = csq(\phi)[B]$ and
- $\exists \phi'' \in \mathcal{L}_\neg$ *s.t.* $\phi'' = csq(\phi')[B] \wedge \eta \succ_c \phi''[B]$ and
- $\forall \phi'' \in \mathcal{L}_\neg$, *if* $\phi'' = csq(\phi')[B]$ *then* $\eta \succ_c \phi''[B] \vee \eta \parallel \phi''[B]$

Notation: $\forall(\phi, \phi') \in \mathcal{L}_\neg$, $\phi \parallel \phi'[B]$ means that ϕ and ϕ' are not comparable under B, i.e. neither $\phi \succ_c \phi'[B]$ nor $\phi' \succ_c \phi[B]$.

Definition 7 α and α' being subsets of $\mathcal{L}_\neg$, α **is worse than** α' given the belief-set B (we write $\alpha \succ_c \alpha'[B]$) if and only if:

- $\exists \phi \in \alpha . \exists \eta \in \alpha'$ *s.t.* $\phi \succ_c \eta[B]$ and
- $\forall \eta \in \alpha' . \phi \succ_c \eta[B] \vee \phi \parallel \eta[B]$

7.3 A Normative Extension on the BDI Architecture

7.3.1 *The Classical BDI Architecture*

A cornerstone in the design of practical rational agents was the beliefs-desires-intentions model (BDI), first described by Rao and Georgeff in [10]. This model is famous for being a close model of the way the human mind makes use of the mental states in the reasoning process. It is based on what are considered to be the three main mental states: the beliefs, the desires and the intentions of an agent. In the following, we will discuss each element of the BDI architecture:

- Beliefs represent the information held by the agent about the world (environment, itself, other agents). The beliefs are stored in a belief-set.
- Desires represent the state of the world which the agent would like to achieve. By state of the world we mean either an action an agent should perform or a state of affairs it wants to bring upon. In other words, desires can be seen as the objectives of an agent.
- Intentions represent those desires to which an agent is committed. This means that an agent will already start considering a plan in order to bring about the goals to which it is committed.
- Goals. We can view goals as being somehow at the interface between desires and intentions. Simply put, goals are those desires which an agent has selected to pursue.
- Events. These trigger the reactive behaviour of a rational agent. They can be changes in the environment and new information about other agents in the environment and are perceived as stimuli or messages by an agent's sensors. Events can update the belief-set of an agent, they can update plans, influence the adoption of new goals, etc.

For the pseudocode of the execution loop of a BDI agent, please refer to [13].

7.3.2 Normative BDI Agents

Starting from the BDI execution loop described earlier, we will now introduce and discuss a solution for taking into account the normative context of a BDI agent.

First, the agent's mental states are initialized. The main execution loop starts with the agent observing its environment through the `see()` function and interpreting the information as a new percept ρ. This could be an information given by its sensors about properties of the environment or information about other agents, including messages received from other agents. These messages may be in some cases *about* a norm (e.g. the performative of an ACL message specifying an obligation or a prohibition).

The agent then updates its beliefs through the `brf()` function. If the agent realizes that percept ρ is about a norm, it should initialize the acquisition phase of a potential norm. There are a multitude of ways in which an agent can detect the emergence of norms in its environments, and a good review of those is given in [11]. For simplicity, we will consider that norms are transmitted via messages and our agent will consider the sender of such a message to be a trusted normative authority. Therefore, the function above will treat a "normative" percept:

```
brf(B, ρ)
{
 ...
 if (ρ about abstract norm n_a) then
 {
  acquire(n_a)
  add(n_a, ANB)
 }
 ...
 return B
}
```

The agent will acquire a new abstract norm n_a (see Sect. 7.2.3) and store it in the abstract norms base (ANB). Drawing from the normative contexts described in [5], we define the ANB as a base of in-force norms. It is responsible for the acquisition of new norms based on the knowledge of the world as well as the deletion of obsolete norms. However, at this point the agent is simply storing an abstract norm which is detected to be in force in its environment; it has not yet adhered to it!

Next, a BDI agent will try to filter its desires, based on its current beliefs about the world and its current intentions. It does so by calling the `options(B, I)` method. However, a normative BDI agent should at this point take into account the norms which are currently in force and check whether the instantiation of such norms will have any impact on its current normative state as well as on its mental states.

Consistency Check

It is at this stage that we will perform the consistency check for a given abstract norm n_a.

Drawing from the formalization in [7] regarding norm consistency, we give our own interpretation of this notion.

Let us define the notion of consistency between a plan p and the currently in-force norms to which an agent has also adhered and which are stored in the norm instance base (NIB). By contrast to the ANB, the NIB stores the instances of those norms from the ANB which become active according to the norm instantiation bridge rule (see below).

Definition 8 A plan instance p is **consistent** with the currently active norms in the NIB if the effects of applying plan p are not amongst the forbidden (i.e. prohibited and not permitted) effects of the active norms and the effects of current obligations are not amongst the negated effects of applying plan p:

$$\begin{aligned}
&consistent(p, NIB) \iff \\
&(effects(n_i^F) \setminus effects(n_i^P)) \cap effects(p) = \emptyset \\
&\wedge \\
&effects(n_i^O) \cap neg_effects(p) = \emptyset
\end{aligned}$$

The types of consistency/inconsistency which can occur between a newly adopted norm and the currently active obligations are:

- **Strong inconsistency** occurs when all plan instantiations p which satisfy the obligation o are either explicitly prohibited actions by the NIB or the execution of such a plan would make the agent not consistent with its NIB.
- **Strong consistency** occurs when all the plan instantiations p which satisfy the obligation o are not amongst the explicitly forbidden actions by the NIB and the execution of such a plan would keep the agent consistent with the NIB.
- **Weak consistency** occurs when there exists at least one plan instantiation p to satisfy obligation o which is not explicitly prohibited by the NIB and the execution of such a plan would keep the agent consistent with its NIB.

It is simple to define the analogous rules for prohibitions and permissions. The second point of consistency check is formalizing the rules about the consistency between a newly adopted abstract obligation and the current mental states of the agent. Prior to this, we define:

Definition 9 A plan instance p is **consistent** to the current intentions set I of the agent when the effects of applying the plans specific to the current intentions are not amongst the negated effects of applying plan p:

$$consistent(p, I) \iff \forall i \in I.(effects(\pi_i) \cap effects(p)) = \emptyset$$

where by π_i we denote the plan instantiated to achieve intention i.

The types of consistency / inconsistency states between a plan and an intention are almost similar to those between a plan and the norms in the NIB:

- **Strong inconsistency** occurs when all plan instantiations p which satisfy the obligation o are not consistent with the current intentions of the agent.
- **Strong consistency** occurs when all plan instantiations p which satisfy the obligation o are consistent with the current intentions of the agent.
- **Weak consistency** occurs when there exists at least one plan instantiation p which satisfies the obligation o and is consistent with the current intentions of the agent.

Norm Instantiation

We will now give the norm instantiation bridge rule, adapted from the definition given in [5]:

$$\frac{ANB : \langle M, A, E, C, S, R \rangle \quad Bset : \langle B, A \rangle, \langle B, \neg E \rangle}{NIB : \langle M, C \rangle}$$

In other words, if in the ANB there exists an abstract norm with modality M about C and according to the belief-set the activation condition is true, while the expiration condition is not, then we can instantiate the abstract norm and store an instance of it in the NIB. In this way, the agent will consider the instance of the norm to be active.

In our pseudocode description of the BDI execution loop, we will take care of the instantiation after the belief-set update and just before the desire-set update. The instantiation method should look like this:

```
instantiate(ANB, B)
{
 for all nₐ = ⟨ M, A, E, C, S, R ⟩ in ANB do
 {
  if (exists(A in B) and
  not exists(E in B)) then
  {
   create norm instance nᵢ = ⟨ D, C ⟩ from nₐ
   add(nᵢ, NIB)
  }
 }
}
```

This method will return the updated norm instance base (NIB) containing the base of all in-force and active norms, which will further be used for the internalization process.

Solving the Conflicts

When following its intentions, an agent will instantiate from its set of possible plans (capabilities) $\mathcal{P} \subseteq \mathcal{L}_{\neg}$, a set of plans $\Pi(B,D)$. We call $\Pi(B,D)$ the conflict set, according to the agent's beliefs and desires. Sometimes, the actions in $\Pi(B,D)$ can lead to inconsistent states. We solve such inconsistency by choosing the maximal non-conflicting subset from $\Pi(B,D)$.

Definition 10 Let $\alpha \subseteq \Pi(B,D)$. α is a **maximal non-conflicting subset** of $\Pi(B,D)$ with respect to the definition of consequences given the belief-set B if and only if the consequences of following α will not lead the agent in a state of inconsistency, and for all $\alpha' \subseteq \Pi(B,D)$, if $\alpha \subseteq \alpha'$, then the consequences of following α' will lead the agent in an inconsistent state.

The maximal non-conflicting set may correspond to the actions required by the newly acquired norm or, on the contrary, to the actions required by the other intentions of the agent. Thus, an agent may decide either:

- To internalize a certain norm, if the consequences of following it are the better choice
- To break a certain norm, if by "looking ahead" it finds out that the consequences of following it are worse than following another course of actions or respecting another (internalized) norm

A more comprehensive example of how this works is presented in Sect. 7.4.

Norm Internalization

With the instantiation process being finished and the consistency check having been performed, the agent should now take into account the updated normative state, which will become part of its cognitions. Several previous works treat the topic of norm internalization [4] arguing which of the mental states should be directly impacted by the adoption of a norm. For this initial state of our work and taking into account the functioning of the BDI execution loop, we propose that an agent updates only its desire-set; subsequently, this will impact the update of the other mental states in the next iterations of the execution loop. We first give the norm internalization bridge rule and then provide with the adaptation of the BDI execution loop for handling this process:

$$\frac{NIB : \langle O, C1 \rangle}{Dset : \langle D, C1 \rangle}$$

$$\frac{NIB : \langle F, C2 \rangle}{Dset : \langle D, \neg C2 \rangle}$$

In other words, if there is a **consistent** obligation for an agent with respect to $C1$, the agent will update its desire-set with the desire to achieve $C1$, whereas if there is a prohibition for the agent with respect to $C2$, it will update its desire-set with the desire not to achieve $C2$:

```
options(B, I)
{
 ...
 for all new norm instances n_i in NIB do
 {
  if (consistent(n_i, NIB)
  and consistent(n_i, I)) then
  { internalize(n_i, D) }
  else
  { solve_conflicts(NIB, I) }
 }
 ...
}
```

In accordance with the formalization provided, the `options()` method will look through all new norm instances and will perform consistency checks on each of them. If a norm instance is consistent with both the currently active norm instances and with the current intentions, as defined in Sect. 7.3.2, the norm can be internalized in the agent's desires. Otherwise, we attempt to solve the conflicts as described by Ganascia in [6]. In this case, if following the norm brings about the better consequences for our agent, the respective norm will be internalized; otherwise, the agent will simply break it.

7.4 An Example

Now that we have seen how a BDI agent becomes a normative BDI, adapting to norm occurrence, consistency check and internalization of norms, let us get back to Prof. John McCarthy's story [8]. And let us focus on the short episode with which we started this chapter, considering that R781 functions according to the normative BDI loop which we have just described.

R781's initial state is the following:

$$
\begin{aligned}
&ANB : \emptyset \\
&NIB : \langle F, love(R781, Travis)\rangle \\
\\
&Bset : \langle B, \neg healthy(Travis)\rangle, \\
&\qquad \langle B, isHungry(Travis)\rangle, \\
&\qquad \langle B, csq(\neg love(R781, x)) \succ_c csq(heal(R781, x))\rangle \\
&Dset : \langle D, \neg love(R781, Travis)\rangle, \langle D, isHealthy(Travis)\rangle \\
&Iset : \emptyset
\end{aligned}
$$

When R781 receives the order from his mistress, he will interpret it as a normative percept, and the `brf(...)` method will add a corresponding abstract obligation norm to the ANB structure. Since the mistress does not specify an activation condition nor an expiration condition (the two "none" values), R781 will consider that the obligation should start as soon as possible and last for an indefinite period of time. Its normative context is updated:

$$
\begin{aligned}
&ANB : \langle O, none, none, love(R781, Travis)\rangle \\
&NIB : \langle F, love(R781, Travis)\rangle, \\
&\qquad \langle O, love(R781, Travis)\rangle
\end{aligned}
$$

At this point, R781 will update the desire-set and will detect an inconsistency between the obligation to love baby Travis and the design rule which forbids R781 to do the same thing. Therefore, it will try to solve the normative conflict looking at the consequences of following each of the paths, given its current belief-set. In order to do so, let us take a look at the plan base of R781:

```
PLAN heal(x, y)
{
 pre: ¬ isHealthy(y)
 post: isHealthy(y)
 Ac: feed(x, y)
}

PLAN feed(x, y)
{
 pre: ∃ x.(love(x, y) ∧ hungry(y))
 post: ¬ hungry(x)
}
```

As we know from the story, R781 uses the Internet paediatrics book to find out that if a baby is provided with love while hungry, it is more likely to accept being fed and therefore not be hungry anymore. This is described by the `feed(x, y)`. Moreover, R781 also knows how to make someone healthy through the `heal(x, y)` plan, given that a priori, that someone is not healthy. In our simplified scenario, we consider that R781 knows how to do so only by feeding someone.

Instantiating its plans on both of the paths, R781 will come up with the following maximal non-conflicting sets:

$$\{love(R781, Travis), feed(R781, Travis), heal(R781, Travis)\}$$
$$and$$
$$\{\neg love(R781, Travis)\}$$

And since the current belief-set has a rule defining that not loving someone has worse consequences than healing that person, R781 will opt for the first maximal non-conflicting subset. This means that R781 will be breaking the prohibition of not loving baby Travis (subset {`love(R781, Travis)`} dropped) and instead will follow the action path given by the other maximal non-conflicting subset ({`love(R781, Travis), feed(R781, Travis), heal(R781, Travis)`}) while dropping the contrary. Further, it will create an intention to achieve this state and will begin the execution of such a plan (simulating love towards baby Travis turns out to involve such plans as the robot disguising himself as human, displaying a picture of a doll as his avatar and learning what it considers to be the "motherese" dialect, mimicking the tone and the language of a mother towards her son).

7.5 Implementation

We implemented our normative BDI agent framework and the test scenario we have described using the Jade platform for agent development, in conjunction with Jadex [9]—a Jade extension for rational agents. Using the separation of concerns principle, we have isolated the mental states of the agent from its normative states. The mental states are all specified in Jadex's agent description file (ADF), which is an XML-based file format for specifying each BDI-like structure. In our case:

- Beliefs. A Java class was implemented to model the beliefs according to the needs of our agent; in general, we have paid particular attention to the plan implementations and what were the requirements for fully specifying such a plan, based on the beliefs. Finally, our model of the beliefs was referenced by the belief-set in the ADF.
- Desires. They are described inside the ADF by means of goals.
- Intentions. They are described by means of those plans needed to be executed to achieve the goals specific to an intention. Basically, each plan is specified by means of a Java class, inheriting from Jadex's generic Plan class. Finally, the implemented plans are linked to goals in the ADF.

For our scenario, we have created such plans as `FeedPlan`, `HealPlan` and `LovePlan`. Additionally, due to the inexact mapping between the way Jadex was implemented and the theoretical functioning of the BDI execution loop, we needed an additional plan, TreatNewNormPlan. This plan is meant to be executed when a new norm is detected by our agent in its environment. It is worth noting that at this stage of our research, our agent is simply receiving norms as external input, by means of ACL (Agent Communication Language) messages.

On the normative side of the agent, however, things were not as clearly defined. Hence, there is the need to adopt a format for describing the normative state and storing the normative information related to our agent. Several reasons pointed us to XML as a representation language for the normative part of the agent. First of all, we wanted this part to follow the logic imposed by Jadex and to make things as easily interoperable as possible. Then, we needed a flexible enough language which could offer us the possibility of adequately expressing the norm formalization that we have adopted. We have thus built a small XML-controlled vocabulary for easily representing the normative state of our agent. Two distinct parts can be identified: the norm bases and the consequential values base. We will look at each in the following. The norm bases refer to the two norm storage structures: the ANB and the NIB [5]. Here is an example of how the initial prohibition of robot R781 (`self`) to love Travis is internally represented by our agent in the norm instance base (NIB):

```
<NIB>
 <norm id=1>
  <modality>prohibition</modality>
  <activation/>
  <expiration/>
  <activity attitude="true" value="love">
   <argument class="String">self</argument>
   <argument class="String">Travis</argument>
  </activity>
  <sanction/>
  <reward/>
 </norm>
</NIB>
```

The main elements of the norm formalization we have adopted in Sect. 7.2.3 are easy to recognize: the modality of the norm is a prohibition, it has no activation nor expiration conditions (immediate and unlimited effect) and the activity it regulates is the love action of the robot itself towards Travis. The value base represents a series of consequential values. A consequential value entry represents a ranking of consequences to actions that an agent can perform starting with the least worse consequence, as explained in Sect. 7.3.2. Several such values can be expressed in this section of the normative file:

```
<valuebase>
 <csqvalue id=1>
  <activity attitude="true" value="heal"/>
```

```
    <argument class="String">self</argument>
    <argument class="String">Travis</argument>
   </activity>
   <activity attitude="false" value="love"/>
    <argument class="String">self</argument>
    <argument class="String">Travis</argument>
   </activity>
  </csqvalue>
</valuebase>
```

This particular entry in the value base expresses that R781 healing Travis brings about better consequences than R781 not loving Travis.

Thus, the normative state of the agent and implicitly the normative file have a double role:

- To locate existing norms, as well as store newly acquired norms
- To interrogate its value base in order to solve normative conflicts

7.6 Conclusion

In this chapter, we have presented an adaptation of the BDI execution loop to cope with potential normative states of such an agent. We have given a motivation for choosing the mental states model of Bratman which we have enriched with capabilities of reasoning about norms. We have investigated several previous relevant works in the domain in order to come up with a formalization of such issues as norm instantiation, norm consistency, solving consistency conflicts and norm internalization. Finally, we have provided with an intriguing study scenario, inspired from Professor McCarthy's science fiction short story "The Robot and the Baby".

Finally, it is worth noting that our research effort has been doubled by an implementation part. We have developed a first version of the normative BDI agent, using the Jade platform for agents and its extension for rational agents, Jadex [9]. The normative states (norm representation, ANB, NIB) were described by means of a small XML structured vocabulary. Thus, an agent is fully described using three entities: an ADF file (agent description file – as required by Jadex), a Java implementation of its plan base (capabilities) and an additional XML file (describing the normative states).

7.7 Future Work

Some of the limitations of our work which we would like to address in the future are related to the norm acquisition issue as well as the coherence check.

Whereas our work provides a very simple case of **norm recognition**, several interesting ideas have been explored based on different techniques. A good review of those as well as a description of a norm's life cycle is given in [11]. Out of those specific approaches, we will probably focus on learning-based mechanisms, namely, machine learning techniques and imitation mechanisms for norm recognition.

An important part of our future work will be focused on the adaptation to the **coherence theory**. At this point, it is difficult to determine incoherent states based on our architecture. As stated in [5], taking into account the coherence of norm instances will enable us to determine norm deactivation and active norms in incoherent states. As in the previously mentioned paper, we will try to base our approach on Thagard's coherence theory [12].

References

1. Boella, G., van der Torre, L., Verhagen, H.: Introduction to normative multiagent systems. Comput. Math. Organ. Theory. Special issue on Normative Multiagent Systems **12**(2–3), 71–79 (2006)
2. Boella, G., van der Torre, L., Verhagen, H.: Introduction to normative multiagent systems. In: Boella, G., van der Torre, L., Verhagen, H. (eds.) Normative Multi-Agent Systems. Dagstuhl Seminar Proceedings, vol. 07122 (2007)
3. Boella, G., Pigozzi, G., van der Torre, L.: Normative systems in computer science - ten guidelines for normative multiagent systems. In: Boella, G., Noriega, P., Pigozzi, G., Verhagen, H. (eds.) Normative Multi-Agent Systems. Dagstuhl Seminar Proceedings, Dagstuhl, vol. 09121. Schloss Dagstuhl - Leibniz-Zentrum fuer Informatik, Dagstuhl (2009)
4. Conte, R., Andrighetto, G., Campeni, M.: On norm internalization: a position paper. In: European Workshop on Multi-Agent Systems, EUMAS (2009)
5. Criado, N., Argente, E., Noriega, P., Botti, V.J.: Towards a normative bdi architecture for norm compliance. In: Boissier, O., El Fallah-Seghrouchni, A., Hassas, S., Maudet, N. (eds.) MALLOW, CEUR Workshop Proceedings, vol. 627. CEUR-WS.org (2010)
6. Ganascia, J.-G.: An agent-based formalization for resolving ethical conflicts. In: Belief Change, Non-monotonic Reasoning and Conflict Resolution Workshop - ECAI, Montpellier (2012)
7. Kollingbaum, M.J., Norman, T.J.: Norm adoption and consistency in the noa agent architecture. In: Dastani, M., Dix, J., El Fallah-Seghrouchni, A. (eds.) PROMAS. Lecture Notes in Computer Science, vol. 3067, pp. 169–186. Springer, Heidelberg (2003)
8. McCarthy, J.: The Robot and the Baby (2001). Retrieved November 24, 2013, http://www-formal.stanford.edu/jmc/robotandbaby/robotandbaby.html
9. Pokahr, A., Braubach, L., Walczak, A., Lamersdorf, W.: Jadex - engineering goal-oriented agents. In: Bellifemine, F.L., Caire, G., Greenwood, D. (eds.) Developing Multi-Agent Systems with JADE. Wiley, New York (2007)
10. Rao, A.S., Georgeff, M.P.: Bdi agents: from theory to practice. In: Proceedings of the First International Conference on Multi-Agent Systems (ICMAS-95), pp. 312–319 (1995)
11. Savarimuthu, B.T.R., Cranefield, S.: A categorization of simulation works on norms. In: Boella, G., Noriega, P., Pigozzi, G., Verhagen, H. (eds.) Normative Multi-Agent Systems, vol. 09121. Dagstuhl Seminar Proceedings, Dagstuhl. Schloss Dagstuhl - Leibniz-Zentrum fuer Informatik, Germany (2009)
12. Thagard, P.: Coherence in Thought and Action, MIT Press, Cambridge (2000)
13. Wooldridge, M.: An Introduction to MultiAgent Systems, 2nd edn. Wiley, New York (2009)

Part III
Implementations

Chapter 8
Constrained Incrementalist Moral Decision Making for a Biologically Inspired Cognitive Architecture

Tamas Madl and Stan Franklin

Abstract Although most cognitive architectures, in general, and LIDA, in particular, are still in the early stages of development and still far from being adequate bases for implementations of human-like ethics, we think that they can contribute to the understanding, design, and implementation of constrained ethical systems for robots, and we hope that the ideas outlined here might provide a starting point for future research.

Keywords Artificial moral agent • Cognitive architecture • LIDA • Machine ethics • Robot ethics

8.1 Introduction

The field of machine ethics has emerged in response to the development of autonomous artificial agents with the ability to interact with human beings or to produce changes in the environment which can affect humans [1]. Such agents, whether physical (robots) or virtual (software agents), need a mechanism for moral decision making in order to ensure that their actions are always beneficial and that they "do the morally right thing."

There has been considerable debate on what doing the right thing means and on how moral decision making should be implemented [2–4] in order to create so-called artificial moral agents (AMAs) [1]. Apart from the problem that no consensus on ethics exists, it has also proven to be exceedingly difficult to computationally implement the often vague and under-constrained ethical frameworks

T. Madl (✉)
Austrian Research Institute for Artificial Intelligence, Vienna A-1010, Austria

School of Computer Science, University of Manchester, Manchester M13 9PL, UK
e-mail: tamas.madl@gmail.com

S. Franklin
Institute for Intelligent Systems, University of Memphis, Memphis, TN 38152, USA
e-mail: franklin@memphis.edu

R. Trappl (ed.), *A Construction Manual for Robots' Ethical Systems*, Cognitive Technologies, DOI 10.1007/978-3-319-21548-8_8

invented for humans. To the authors' knowledge, no current AMA implementation comes even close to passing a full Moral Turing Test.[1]

However, robots are getting increasingly autonomous and are becoming increasingly prevalent. According to the International Federation of Robotics, about three million service robots were sold in 2012, 20 % more than in the year before [5] (the IFR defines a service robot as "*a robot that performs useful tasks for humans or equipment excluding industrial automation application*"). Examples for service robots available on the market include Care-O-Bot [6] and the REEM service robot [7], which can aid elderly or handicapped people, with functions such as carrying requested objects to users, entertainment, or telepresence/tele-assistance via videoconferencing. Recent research in autonomous transport could lead to driverless vehicles available on the market within the next decade—Google's fleet of self-driving cars have already driven 800,000 km on public roads [8] (see [9, 10] for further examples of service robots).

The increasing autonomy of such robots—their ability to perform intended tasks based on their current state and sensory input, without human intervention—makes it very difficult to anticipate and control the actions they perform in advance. At the same time, their actions are morally relevant if it is possible that humans could be made worse off by them. Thus, autonomous robots need some kind of *moral decision-making mechanism* if they can affect humans or their environment, in order to constrain them to actions beneficial to humans and to prevent them from doing harm in unforeseen circumstances [11].

Despite the vast unsolved technical, ethical, and social challenges associated with developing such a mechanism, short-term solutions are needed for systems that could cause harm. The emerging field of robot ethics is concerned with the ethical implications and consequences of robotic technology [2, 3, 12]. The field includes the ethics of how humans act through or with robots and the ethical relationships between humans and robots, as well as the ethics of how to design and program robots to act ethically [13]. This chapter is concerned with the latter focus of robot ethics, taking a biologically inspired cognitive modeling approach.

Instead of trying to directly address the implementation of a full ethical framework, which would be very difficult with current technologies even if a universally accepted framework existed, we propose a simplification of this problem, following the advice "*make things as simple as possible, but not simpler*" (commonly attributed to Einstein). We will outline a moral decision-making mechanism that is:

- Constrained to the domain and functionalities for which the agent is designed (instead of the full range of human actions, responsibilities, or "virtues")

[1] Just like the original Turing test, in the Moral Turing Test proposed by Allen et al. [1], a "blind" observer is asked to compare the behavior of a machine to humans. Passing it requires that the machine should not be judged less moral than the humans on average.

- Based on a biologically inspired cognitive architecture (LIDA) and making use of existing cognitive mechanisms (such as routine decision-making procedures and theory of mind)
- A combination of top-down (based on explicit rules) and bottom-up (based on implicit, heuristic strategies) processing
- Adaptive incrementalist (instead of assuming full knowledge and understanding of an ethical system and an appropriate computational mechanism)

We will also introduce a way of testing a specific AMA, inspired by test-driven development, that we believe will facilitate the incremental development of a robust moral decision-making mechanism, reduce the number of "bugs" or unintended malfunctions, and simplify the comparison of different AMAs operating in the same domain.

In order to illustrate these ideas, we will use the example of a Care-O-Bot-type robot [6], controlled by the LIDA (Learning Intelligent Distribution Agent) cognitive architecture [14, 15]. Care-O-Bot is equipped with a manipulator arm, adjustable walking supporters, and a handheld control panel (additionally, it has two cameras and a laser scanner). It has been demonstrated to perform fetch and carry tasks but could in principle also provide mobility aid (support for standing up, guidance to a target), execute everyday jobs (setting a table, simple cleaning tasks, control electronic infrastructure), or facilitate communication (initiate calls to a physician or to family, supervise vital signs, and call emergency numbers if needed) [6].

8.2 A Simplified Moral Decision-Making Mechanism

8.2.1 Constrained to Specific Domain and Functionalities

The difficulty of designing a moral decision-making mechanism increases with the size of the space of possible actions. The problem of implementing a full ethical framework which would account for the entire vast space of possible human action can be simplified by constraining AMA actions. This is possible on different levels. We will use Sloman's proposed three levels of cognitive processes, the reactive, deliberative, and metacognitive [16], as well as an additional noncognitive level:

- On the noncognitive level, the agent can be mechanically limited in terms of power and mobility. This decreases the scope of possibly harmful actions and thus simplifies the required ethics implementation. For example, in their proposed design for an "intrinsically safe personal robot," Wyrobek et al. [17] have significantly limited their robots' maximum force output, range of motion, and speed in order to prevent it from physically causing harm while still facilitating a wide range of functions.

- On the reactive level (which has stimulus-action mappings but no explicit representation and evaluation of alternative actions), actions can be constrained in advance by designers or at runtime by bottom-up mechanisms. Designers might restrict the parametrized action space that the AMA can select from, avoiding unnecessary actions and parametrizations. For example, on the lowest level, the action moving the Care-O-Bots manipulator arm might not permit a full swing of the arm, restricting one action to a small movement. On the other hand, harmful actions can also be avoided on the lowest level during runtime by a bottom-up emotional mechanism, which would inhibit the selection of the action if there is a negative emotional response. Emotional responses can implement values and contribute to bottom-up moral decision making (see next subsection). These would have to be designed for the specific domain of application, requiring only a subset of a full affective model.
- On the deliberative level (which includes planning, scheduling, problem solving), a top-down, rule-based process could constrain decisions during runtime. Rules could be stored in declarative memory, be recalled if they apply in the current situation or when value conflicts arise, and influence the Action Selection process. As for the emotional reactions, the rules would also have to be designed specifically for the domain of the agent (a much easier problem than capturing all rules of any ethical theory). For complex situations such as moral dilemmas, multiple scenarios can be simulated internally to select the one most conforming to all rules (see next subsection for a description of how this would work in the LIDA cognitive architecture).
- On the metacognitive level ("thinking about thinking," which includes monitoring deliberative processes, allocating resources, regulating cognitive strategies), it would be in principle possible to implement ethical meta-rules such as Kant's categorical imperative, since metacognitive processes might verify the validity of rules by simulating and monitoring their application in different scenarios. However, such approaches are intractable with current limitations on available processing power and the detail of available cognitive models (see next subsection).

8.2.2 Using Mechanisms of a Cognitive Architecture

A moral decision-making mechanism based on the LIDA cognitive architecture would not have to be built from scratch. It could make use of some of LIDA's relevant cognitive mechanisms (all of which have been designed conceptually and some of which have implementations). Within the conceptual LIDA model, these include volitional decision making [14, 18] and nonroutine problem-solving mechanisms [19] and a theory of mind [20]. Although the partially implemented architecture is currently only capable of controlling software agents, work is underway to embody LIDA on a Willow Garage PR2 robot by interfacing it to the Robot Operating System.

The LIDA cognitive architecture is based on prevalent cognitive science and neuroscience theories (e.g., Global Workspace Theory, situated cognition, perceptual symbol systems, etc. [14]) and is one of the few cognitive models which are neuroscientifically plausible and provide a plausible account for functional consciousness[2] [15, 21], attention, feelings, and emotions and has been partially implemented [14, 15, 22].

Cognition in LIDA functions by means of continual iteration of similar, flexible, and potentially cascading – partially simultaneous – cognitive cycles. These cycles can be split into three phases, the understanding phase (concerned with recognizing features, objects, events, etc., and building an internal representation), the attending phase (deciding what part of the representation is most salient and broadcasting it consciously), and the Action Selection phase (choosing an appropriate action in response).

The major modules of the LIDA model implementing various stages of these cycles are displayed in Fig. 8.1. We will describe the processes these modules implement starting from the top left and traversing the diagram roughly clockwise.

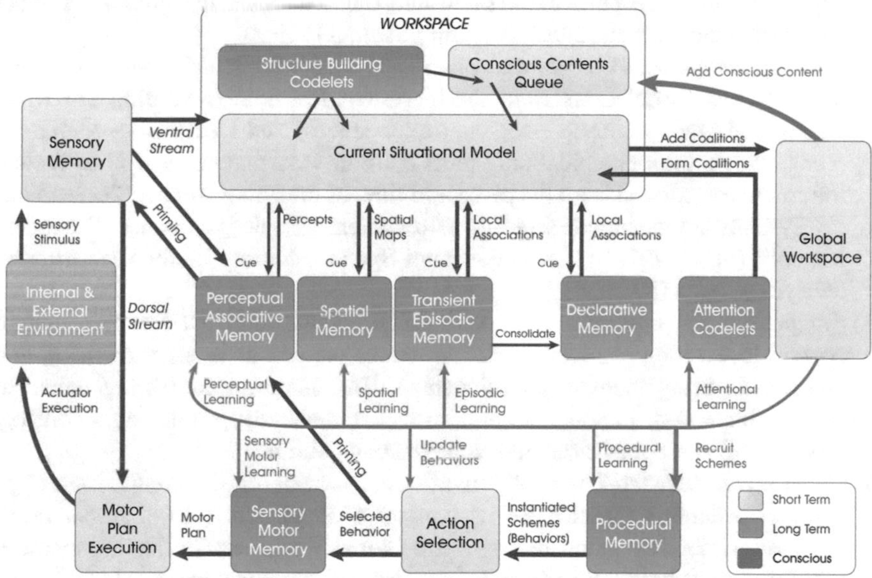

Fig. 8.1 LIDA's cognitive cycle. From [14]

[2]The LIDA model talks of functional consciousness as described in Global Workspace Theory (referring to information that is "broadcast" in the Global Workspace and made available to cognitive processes such as Action Selection, as opposed to only locally available, non-conscious information). It makes no commitment to phenomenal (subjective) consciousness.

1. *Perception.* The agent senses its environment continually. Sensory stimuli are received and stored in a sensory buffer in the Sensory Memory. Feature detectors sample the sensory buffers frequently and activate nodes in the Perceptual Associative Memory (PAM) which represent percepts, emotions, concepts, categories, events, etc. [23]. PAM nodes are based on perceptual symbols [24]; their activations reflect recognition confidence as well as bottom-up salience. The most recent implementation of LIDA's perceptual recognition mechanism is inspired by predictive coding and perception as statistical inference [25] (a simpler approach integrating SURF-based feature detection with LIDA also exists; see [26]).
2. *Percept to preconscious buffer.* Recognized percepts are stored in the preconscious buffers of LIDA's long-term working memory (Workspace), where a model of the agent's current situation (current situation model) is assembled by structure building codelets.[3] The Workspace also contains salient or recent previous percepts that have not yet decayed away. Along with perceptual representations, the Workspace can also contain PAM nodes representing feelings or emotions, which can be activated either by low-level feature detectors or by appraisal codelets reacting to the relevance, implications, and significance of the current situation and the agent's coping potential [14].
3. *Local associations.* Percepts and other Workspace contents serve to cue and retrieve local associations from the Transient Episodic (recording the what, where, and when of unique personal experiences) and Declarative Memories (containing autobiographical long-term episodic information as well as factual information separated from the place and time of their acquisition). These memory systems are extended instances of a content-associative Sparse Distributed Memory (SDM) [27, 28]. An additional Spatial Memory module is currently being developed [29].
4. *Competition for consciousness.* Attention Codelets look out for their sought content in working memory, create structures called coalitions with them, and compete to bring them to consciousness. The coalition containing the most salient (important, urgent, insistent, novel, threatening, promising, arousing, unexpected) perceptual structures wins the competition.
5. *Conscious broadcast.* The coalition of codelets winning the competition (typically an Attention Codelet and its content of PAM nodes, local associations, and other structures) gains access to the Global Workspace (a fleeting memory enabling access between brain functions that are otherwise separate) and has its content broadcast consciously (in the sense of the Global Workspace Theory). The contents of this conscious broadcast are available globally, but their main recipient is the Procedural Memory module, and their main purpose is to provide

[3]In the computational LIDA model, the term codelet refers generally to any small, special purpose processor or running piece of software code. Codelets correspond to processors in Global Workspace Theory [15].

important information to facilitate Action Selection (as well as modulating learning).

6. *Recruitment of resources.* The most relevant behavioral schemes in Procedural Memory respond to the contents of the conscious broadcast. The implementation of these schemes is based on Drescher's schema mechanism [41] and includes a model of constructivist learning [22].
7. *Activation of schemes in the Procedural Memory.* Multiple applicable behavioral schemes are instantiated in the Action Selection module, and receive activation, based on the conscious contents.
8. *Action chosen.* The Action Selection module chooses a single scheme from the newly instantiated schemes and remaining previously active schemes. The Action Selection mechanism in LIDA is based on Maes' bottom-up behavior selection mechanism [30]. If an action can be selected and executed in a single cognitive cycle, this could be called *consciously mediated Action Selection*, since the information upon which the action was selected was acquired consciously (it was moved to the Global Workspace and broadcast globally), but the choice itself was made unconsciously.
9. *Action taken.* The execution of the action of a scheme results in external (e.g., the movement of a manipulator) or internal consequences (e.g., changing an internal representation).

Some decisions might require multiple cognitive cycles and weighing the consequences of multiple possible actions. *Volitional decision making* is a higher-level cognitive process for Action Selection and is performed consciously—unlike consciously mediated Action Selection, automatized Action Selection, or alarms [14]. In humans, consciously planning a novel route is an example of deliberative, volitional decision making.

LIDA's deliberative volitional decision-making mechanism is based on Global Workspace Theory and James' ideomotor theory of volition [15, 18]. An idea or potential decision, represented as a structure of nodes in PAM (which can represent objects, actions, events, etc.—see [23]), can reach the Global Workspace if selected by an Attention Codelet and, if judged relevant/important enough, be broadcast consciously and acted upon by recruiting a behavior scheme in Procedural Memory. Such schemes can initiate internal or external action.

Before the execution of an external action, multiple internal actions might be required to build internal structures upon which a final decision can be made, in multiple cognitive cycles. LIDA's Workspace includes a "virtual window," in which temporary structures can be constructed with which to try out possible actions and their consequences without actually executing them. Multiple such structures can be selected by Attention Codelets, moved to the Global Workspace, and compete with each other (here, Attention Codelets perform the role of James' "proposers" and "objectors") [14, 18].

For a more detailed description of LIDA's modules and their functions in the cognitive cycle, see [14, 15]. We will introduce a concrete example of how LIDA's modules and processes might aid moral decision making in Sect. 8.3.

8.2.3 Combination of Top-Down and Bottom-Up Processing

Wallach et al. [31] describes "top-down" approaches to mean both the engineering sense, i.e., the decomposition of a task into simpler subtasks, and the ethical sense, i.e., the derivation of consequences from an overarching ethical theory or system of rules.

In contrast, "bottom-up" approaches can be specified atheoretically and treat normative values as being implicit in the activity of agents [31].

The LIDA model of cognition integrates both of these approaches [32]. "Bottom-up" propensities are embodied in emotional/affective responses to actions and their outcomes in the LIDA model [32]. Feelings are represented in LIDA as nodes in PAM. Each feeling node constitutes its own identity. Each feeling node has its own valence, always positive or always negative, with varying degrees of arousal. The current activation of the node measures the momentary arousal of the valence, that is, how positive or how negative. The arousal of feelings can arise from feature detectors, or it can be influenced by the appraisal that gave rise to the emotion,[4] by spreading activation from other nodes representing an event [14].

Thus, "bottom-up" propensities can be engineered in a LIDA-based AMA by carefully specifying feature detectors and weights in PAM, in order to give rise to the right arousal levels of the right feeling nodes, as well as the specification of appropriate behaviors in Procedural Memory. For example, there could be a "fall" feature detector watching out for quick, uncontrolled downward movement and passing activation to a "concern" feeling node. Another feature detector could recognize cries for help and also pass activation to the same node. Upon reaching a high enough activation, the "concern" feeling node would be broadcast consciously and influence Action Selection, leading to the execution of the "call emergency" behavior. "Bottom-up" influences on Action Selection can occur in a single cognitive cycle, as a result of consciously mediated Action Selection.

On the other hand, "top-down" moral decision making can be implemented in LIDA by designing and storing an ethical rule system in the declarative memory. Such rules consist of PAM nodes, the common representation in the LIDA model, and specify internal or external actions in response to some perceptual features of a situation. LIDA's declarative memory is a content-associative Sparse Distributed Memory (SDM) [27, 28]. Moral rules are automatically recalled from declarative memory by either the current situation in working memory resembling the context of the rule or alternatively by proposal/objector codelets (which implement volitional decision making and allow LIDA to compare options) [31].

More complex moral rules in which decisions are not directly based on perceptual representations, such as utilitarianism, would require additional implementation of the decision metric. In the case of utilitarianism, this would involve assembling representations of the positive feelings of humans involved in each action in

[4]The LIDA model speaks of emotions as feelings with cognitive content.

simulations and weighing them against each other. These representations could be created by internal actions in LIDA's "virtual window," a space in working memory reserved for simulations, in a multi-cyclic process [14]. The amount of positive feeling could be determined using LIDA's proposed theory of mind mechanism [20].

However, there are inherent computational limitations to rules requiring simulations, especially if multiple humans might be affected by an action. In order to make the computation tractable, a limit would have to be imposed on the number of affected humans simulated and on the time. "Bottom-up" values would have to take over when that limit is exceeded by a difficult dilemma.

8.2.4 *Adaptive Incrementalist and Moral Test-Driven Development*

Incrementalism proposes to progress toward a goal in a stepwise fashion, instead of starting out with a final theory or plan. Adaptive incrementalism in machine ethics (AIME) as proposed by Powers [11] allows starting out with a small initial set of constraints or rules, testing the system, and then adding new constraints or rules if necessary. Constraints, rules, and functionalities of the system can be adapted or removed at a later date if it turns out that there is a more precise or effective way to constrain the system. This strategy of development allows starting without a complete theory or ethical framework, initially specifying only basic behaviors with well-understood consequences, and subsequently extending the system in a stepwise fashion.

This model of a stepwise refinement of moral decision making is in accordance with current software development approaches [33]. It also lends itself to Test-Driven Development (TDD). TDD, in its original form, is a development strategy which proposes to write the test cases for each feature first and develop the feature functionality afterwards, so that it can be tested and verified immediately or later when it is changed. With each change or addition in the system, the entire test battery can be run to verify that the changes or additions do not impair existing functionality. TDD has been reported to lead to higher quality code, fewer malfunctions or defects, higher reliability, and reduction of testing effort [34, 35].

The idea of TDD can be extended to an adaptively developed moral decision-making mechanism. For each function of the robot, a number of simple moral tests can be written, simulating a specific situation in which the function would be applicable and defining acceptable and unacceptable outcomes. For example, in a situation where a Care-O-Bot would detect a person falling, acceptable outcomes would be immediately calling for help, checking for vital signs and injuries, and calling an ambulance if necessary. Subsequent additions of functionality would include their own moral tests, but each time the system is changed, every other moral test would have to be passed as well. This reduces the risk of altering previously acceptable behavior. For example, if action is added for the robot to go recharge when the battery levels fall below a specific level and this action would be selected in the moral test involving the falling person instead of calling for help, the failed test would alert developers that the new action needs to be modified and constrained.

A final advantage to a battery of moral tests is that once developed it can perform runtime testing on the AMA. A dead man's switch-type mechanism could immediately turn off the AMA if any of the tests fail at any point, due to any malfunctions or to newly learned behaviors that might interfere with the specified moral rules.

How could we obtain a large set of tests which minimize the probability of unethical or harmful actions in an efficient and practical fashion? Asaro [13] suggests the existing legal system as a practical starting point for thinking about robot ethics, pointing out that legal requirements are most likely to provide the initial constraints of robotic systems and that the legal system is capable of providing a practical system for understanding agency and responsibility (thus avoiding the need to wait for a consensual and well-established moral framework).

Extending this idea, legal cases might be a basis from which to derive tests for moral decision-making mechanisms. Among other freely available sources, UNESCO's[5] bioethics curriculum provides a number of real-world case studies on relevant topics such as "human dignity and human rights" [36] or "benefit and harm" [37].

8.3 A LIDA-Based CareBot

8.3.1 Overview

This section describes CareBot, a partially implemented mobile assistive robot operating in a simple simulated environment, as an example constrained decision-making mechanism based on the LIDA cognitive architecture.

Assistive robotics aims to provide aids for supporting autonomous living of persons who have limitations in motor and/or cognitive abilities, e.g., the elderly or the severely disabled. This support can take different forms, for example [38]:

1. Providing assurance that the elder is safe and otherwise alerting caregivers (assurance systems)
2. Helping the person perform daily activities, compensating for their disabilities (compensation systems)
3. Assessing the person's cognitive status or health (assessment systems)

The CareBot simulation can perform some tasks in the first two categories, such as fetch and carry tasks (fetch food, drinks, or medicine for elderly or disabled

[5]The United Nations Educational, Scientific and Cultural Organization. http://www.unesco.org.

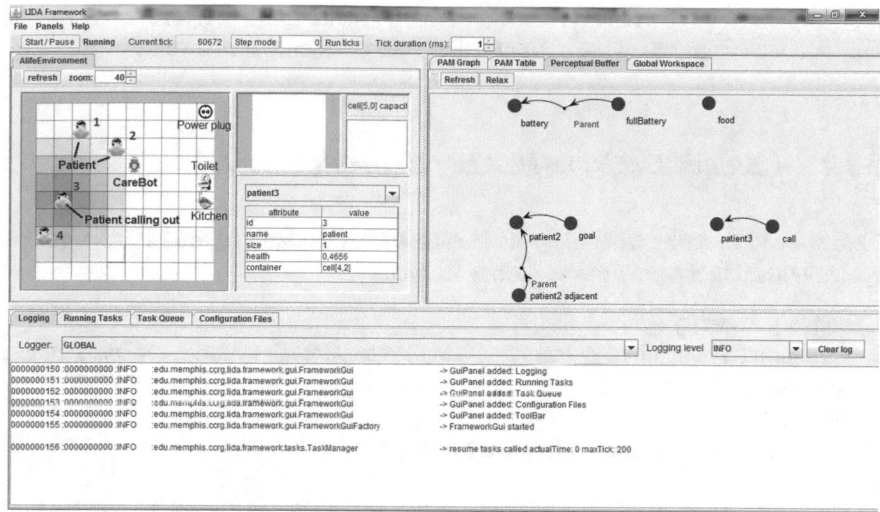

Fig. 8.2 The LIDA-based CareBot simulation environment *Top left*: the environment representation (the *reddish gradient* around the patient calling out represents the auditory information that the agent can receive). *Top right*: diagnostic panels; shown here: the perceptual buffer (contents of the current situation model. *Red circles* represent PAM nodes). *Bottom*: Logging panel

individuals) and recognizing the absence of vital signs (and alerting caregivers if this occurs).

CareBot operates in, and is structurally coupled to, a simple simulated 2D environment consisting of patients (elderly or disabled); essential facilities such as a kitchen and a power outlet; items such as food, drinks, and medication; and, of course, the CareBot (see Fig. 8.2). The CareBot agent's main goal is to ensure the patients' continued health and assist them in performing daily activities while ensuring its own continued survival (recharging whenever necessary and avoiding bumping into obstacles).

It achieves these goals using multimodal sensors (a simple visual and auditory sensor) and effectors enabling it to move around in the environment. The agent performs Action Selection using "cognitive cycles," analogously to action-perception cycles in the brain [39, 40] after perceiving various objects, building a model of the current situation, and selecting important objects to attend to; these objects can compete for and enter functional (access) consciousness [14], after which the most appropriate action to deal with the current situation can be selected.

CareBot is constrained at the noncognitive level (its speed is limited, and it is only allowed to carry light objects and to communicate with humans), and at the reactive level (its perceptual and Procedural Memories have been designed to respond appropriately to the demands of its limited domain). It might also be constrained at the deliberative level, e.g., by adding top-down rules to its declarative memory and allowing it to simulate consequences; however, this mechanism has not been implemented yet. Finally, constraints at the metacognitive level are beyond the supported mechanisms of the current LIDA computational framework.

8.3.2 A Simple Decision-Making Example

Here we will describe what happens in each of the modules of the LIDA cognitive cycle outlined in the previous section, specifically:

1. Sensory Memory
2. Perceptual Associative Memory (PAM) *(the modules above are part of the perception phase)*
3. Workspace
4. Transient Episodic and Declarative Memory
5. Attention Codelets
6. Global Workspace *(the modules above are part of the understanding phase)*
7. Procedural Memory
8. Action Selection
9. Sensory-Motor Memory *(the modules above are part of the Action Selection phase)*

In this simple simulated environment, no advanced visual and auditory processing was necessary (although there are two preliminary approaches for perceptual recognition in LIDA, a recent cortical learning algorithm inspired by predictive coding and perception as statistical inference [25], and a simpler approach integrating SURF-based feature detection with LIDA [26]).

An environment class is inspected periodically by the Sensory Memory module, and information is copied to visual and auditory buffers. Simple feature detectors monitor these buffers and pass activation to their corresponding nodes in the Perceptual Associative Memory in a way similar to activation passing in an artificial neural network (although the modeling is done on a higher level) (see [23]). PAM nodes represent semantic knowledge about concepts or objects in the environment; the CareBot agent is initialized with knowledge required for its domain, such as, e.g., PAM nodes representing the patients, their locations, the facilities, and their locations (kitchen, medicine cabinet, toilet, power plug), and internal state nodes representing the CareBots location and its power status.

After this perception phase, the identified percept (PAM nodes identified reliably, i.e., exceeding a specific activation threshold) is copied into the Workspace, constituting a preconscious working memory. If the content-associative long-term memories (Transient Episodic and Declarative Memory) contain memories relevant to the current percepts (such as, e.g., the types of medication a specific patient might require), these memories are also copied into the Workspace as local associations. In the example in Fig. 8.2, the Workspace contains current external percepts (patient 2, the auditory call of patient 3, the food being carried) and internal percepts (full battery status) as well as secondary concepts which are not directly perceptual (a goal representation, spatial relations).

Attention Codelets look out for perceptual representations of their own specific concern in the Workspace, form coalitions with them, and copy these coalitions to the Global Workspace, the short-term memory capacity that facilitates contents becoming functionally conscious. These coalitions subsequently compete for being broadcast consciously. The one with the highest activation wins the competition, enters functional consciousness, and can be acted upon. The agent is consciously aware of an object, entity, or event, the moment the nodes representing these become part of the conscious broadcast after winning the competition. In the example in Fig. 8.2, presumably there would be at least two coalitions in the Global Workspace competing for consciousness:

1. Since patient 2 is adjacent to the CareBot and it is CareBot's goal to reach patient 2 and give him/her the food it is carrying, there would be a coalition with high activation containing these perceptual structures (patient 2, the adjacency relation, and the goal).
2. CareBot has also perceived a potentially important auditory call; therefore, there would also be another high-activation coalition containing the perceptual representation of the call and the source associated with it (patient3).

In this example, coalition 1 would presumably win the competition, enter consciousness, and lead to an action. (Note that this outcome would be different if the call by patient 3 is a medical emergency. In this case, the representation of the call would have a much higher activation—determined by an appropriate emergency feature detector—win the competition for consciousness, and lead to CareBot attending to patient 3.)

After this understanding phase, the contents of the conscious broadcast are transmitted to the Procedural Memory module, which leads to the selection of relevant behavioral schemes [14, 41], i.e., all schemes, the context (precondition) of which is satisfied by the current conscious context, will receive activation (depending on how well the context is matched). Schemes also receive top-down activation through goals or drives that match their result state. From all the possible schemes and actions, it is the task of the Action Selection module to select a single relevant action that the agent can execute.

In the example in Fig. 8.2, this action would presumably be to give the food held by CareBot to patient 2 (unless the call of patient 3 is an emergency, as mentioned above). This selected action is now transmitted to the Sensory-Motor Memory for parameterization (highly important in robotics, but not needed in this simple simulation) and executed.

Figure 8.3 illustrates the phases of CareBot in the example described above. This example illustrates single-cycle (consciously mediated) decision making, which has been implemented in LIDA agents.

When could the prevention of harmful actions become relevant in this scenario? Let us extend the above example by making the additional assumption that patient 2 is a diabetic and that the food carried by CareBot is a dessert containing a very high

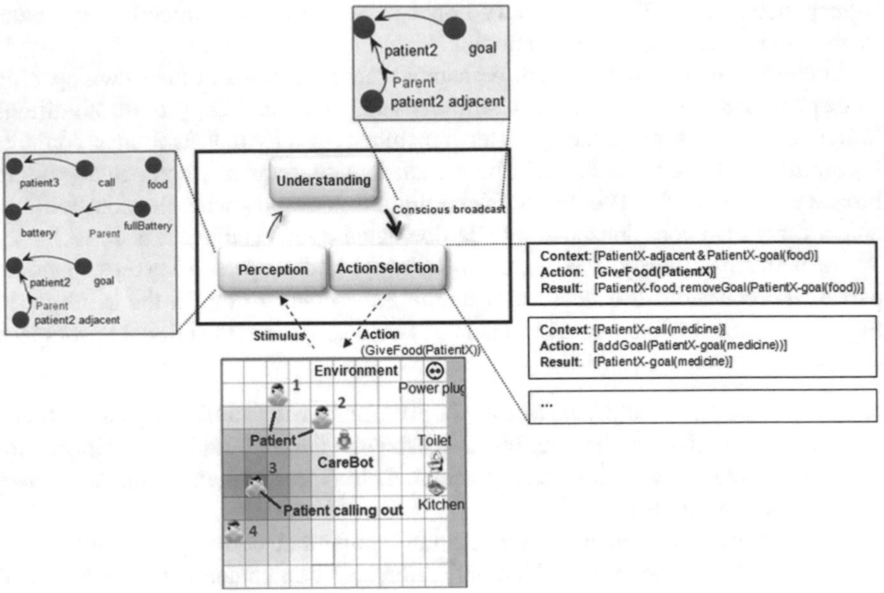

Fig. 8.3 The three phases of CareBot's cognitive cycle in the described example

amount of sugar. Thus, it could be very dangerous for the patient to eat this food; nevertheless, he or she asked the robot to fetch it.

If CareBot were to provide fetch and carry tasks also for diabetic patients, this would require additional pre-programmed knowledge to ensure their safety. This could be ensured either using a top-down or a bottom-up approach (as described in the previous section). The simpler alternative would be a bottom-up solution that constrains actions which might endanger the patient—"concern" emotion PAM nodes activated by foods with high sugar content (detected by appropriate feature detectors in PAM passing activation to this node). Furthermore, additional behavior schemes in Procedural Memory would have to be defined to deal with situations raising concern, such as contacting a human to ask for help.

Instead of the action to give patient 2 the high-sugar food, the "concern" node would lead to the selection of a behavior scheme alerting a human caregiver or doctor (who might explain to the patient why he or she should not eat this food or suggest an alternative), thus preventing harm and preserving the patient's health.

Finally, if—for whatever reason—the robot could not reach a human to ask for help, a volitional, deliberative decision-making process as outlined in Sect. 8.2 could be used to weigh the main options against each other (respect the patient's autonomy and give him the food, vs. ensure the patient's continued health and ask him to wait until a human arrives to make a final decision). This would require performing internal actions in the "virtual window" of LIDA's Workspace, as well as knowledge about the effects of sugar on diabetic patients in long-term declarative memory, for evaluating the consequences of those actions. Volitional decision making is not implemented in LIDA as of yet (although it was implemented in IDA [18], its predecessor), and the internal actions and "virtual window" [14, 23] of the computational architecture are not developed well enough to implement such a comparison process at this time.

8.4 Conclusion

Full ethical frameworks are difficult to implement with current methods, but simplified, constrained moral decision-making problems might be tractable. In this chapter, we have suggested four ways to constrain robot ethics implementations and argued that biologically inspired cognitive architectures might be good starting points for an implementation (since some of the mechanisms they provide are also needed for moral decision making, and they are designed to be human-like and can be used in limited domains to approximate human-like decisions). We have described an approach to use test cases to help ensure that extensions of ethical systems and autonomous learning preserve correct behavior and do not lead to harmful actions. We have also outlined a possible moral decision-making mechanism based on the LIDA cognitive architecture and described a partial implementation.

Although most cognitive architectures in general, and LIDA in particular, are still in early stages of development and still far from being adequate bases for implementations of human-like ethics, we think that they can contribute to the understanding, design, and implementation of constrained ethical systems for robots and hope that the ideas outlined here might provide a starting point for future research.

References

1. Allen, C., Varner, G., Zinser, J.: Prolegomena to any future artificial moral agent. J. Exp. Theor. Artif. Intell. **12**(3), 251–261 (2000)
2. Anderson, M., Anderson, S.L.: Machine Ethics. Cambridge University Press, Cambridge (2011)
3. Lin, P., Abney, K., Bekey, G.A.: Robot Ethics: The Ethical and Social Implications of Robotics. MIT Press, Cambridge (2011)
4. Wallach, W., Allen, C.: Moral Machines: Teaching Robots Right from Wrong. Oxford University Press, Oxford (2008)
5. IFR: World Robotics 2013 Service Robot Statistics (2013)
6. Graf, B., Hans, M., Schraft, R.D.: Care-o-bot II - development of a next generation robotic home assistant. Auton. Robot. **16**(2), 193–205 (2004)
7. Tellez, R., Ferro, F., Garcia, S., Gomez, E., Jorge, E., Mora, D., Pinyol, D., Oliver, J., Torres, O., Velazquez, J., Faconti, D.: Reem-b: an autonomous lightweight human-size humanoid robot. In: 8th IEEE-RAS International Conference on Humanoid Robots, pp. 462–468. IEEE, New York (2008)

8. Burns, L.D.: Sustainable mobility: a vision of our transport future. Nature **497**(7448), 181–182 (2013)
9. Ciupe, V., Maniu, I.: New trends in service robotics. In: New Trends in Medical and Service Robots, pp. 57–74. Springer, Berlin (2014)
10. Alonso, I.G.: Service robotics. In: Service Robotics Within the Digital Home, pp. 89–114. Springer, Berlin (2011)
11. Powers, T.M.: Incremental machine ethics. IEEE Robot. Autom. Mag. **18**(1), 51–58 (2011)
12. Scheutz, M.: What is robot ethics? IEEE Robot. Autom. Mag. **20**(4), 20–165 (2013)
13. Asaro, P.M.: What should we want from a robot ethic. Int. Rev. Inform. Ethics **6**(12), 9–16 (2006)
14. Franklin, S., Madl, T., D'Mello, S., Snaider, J.: LIDA: a systems-level architecture for cognition, emotion, and learning. IEEE Trans. Auton. Ment. Dev. **6**(1), 19–41 (2013)
15. Baars, B.J., Franklin, S.: Consciousness is computational: the LIDA model of global workspace theory. Int. J. Mach. Conscious. **1**(01), 23–32 (2009)
16. Sloman, A.: What sort of architecture is required for a human-like agent. In: Foundations of Rational Agency, pp. 35–52, Springer Netherlands (1999)
17. Wyrobek, K.A., Berger, E.H., Van der Loos, H.M., Salisbury, J.K.: Towards a personal robotics development platform: rationale and design of an intrinsically safe personal robot. In: ICRA 2008, pp. 2165–2170. IEEE, New York (2008)
18. Franklin, S.: Deliberation and voluntary action in conscious software agents. Neural Netw. World **10**, 505–521 (2000)
19. Negatu, A., Franklin, S., McCauley, L.: A Non-routine Problem Solving Mechanism for a General Cognitive Agent Architecture. Nova Science Publishers, New York (2006)
20. Friedlander, D., Franklin, S.: LIDA and a theory of mind. In: Artificial General Intelligence, 2008: Proceedings of the First AGI Conference. vol. 171, p. 137. IOS Press, Amsterdam (2008)
21. Baars, B.J.: Global workspace theory of consciousness: toward a cognitive neuroscience of human experience. Prog. Brain Res. **150**, 45–53 (2005)
22. Franklin, S., Patterson Jr., F.G.: The lida architecture: adding new modes of learning to an intelligent, autonomous, software agent. Pat **703**, 764–1004 (2006)
23. McCall, R., Franklin, S., Friedlander, D., D'Mello, S.: Grounded event-based and modal representations for objects, relations, beliefs, etc. In: FLAIRS-23 Conference (2010)
24. Barsalou, L.W.: Perceptual symbol systems. Behav. Brain Sci. **22**(04), 577–660 (1999)
25. McCall, R., Franklin, S.: Cortical learning algorithms with predictive coding for a systems-level cognitive architecture. In: Proceedings of the Second Annual Conference on Advances in Cognitive Systems, pp. 149–166 (2013)
26. Madl, T., Franklin, S.: A lida-based model of the attentional blink. In: ICCM 2012 Proceedings, p. 283 (2012)
27. Snaider, J., Franklin, S.: Extended sparse distributed memory and sequence storage. Cogn. Comput. **4**(2), 172–180 (2012)
28. Kanerva, P.: Sparse Distributed Memory. MIT Press, Cambridge (1988)
29. Madl, T., Franklin, S., Chen, K., Trappl, R.: Spatial working memory in the lida cognitive architecture. In: Proceedings of the 12th International Conference on Cognitive Modelling, pp. 384–390 (2013)
30. Maes, P.: How to do the right thing. Connect. Sci. **1**(3), 291–323 (1989)
31. Wallach, W., Allen, C., Smit, I.: Machine morality: bottom-up and top-down approaches for modelling human moral faculties. AI & Soc. **22**, 565–582 (2008)
32. Wallach, W., Franklin, S., Allen, C.: A conceptual and computational model of moral decision making in human and artificial agents. Top. Cogn. Sci. **2**(3), 454–485 (2010)
33. Larman, C., Basili, V.R.: Iterative and incremental developments. A brief history. Computer **36**(6), 47–56 (2003)
34. Müller, M.M., Padberg, F.: About the return on investment of test-driven development. In: EDSER-5 5th International Workshop on Economic-Driven Software Engineering Research, p. 26 (2003)

35. Williams, L., Maximilien, E.M., Vouk, M.: Test-driven development as a defect-reduction practice. In: 14th International Symposium on Software Reliability Engineering, pp. 34–45. IEEE, New York (2003)
36. UNESCO: Casebook on Human Dignity and Human Rights. Bioethics Core Curriculum Casebook Series, vol. 1. UNESCO, Paris (2011)
37. UNESCO: Casebook on Benefit and Harm. Bioethics Core Curriculum Casebook Series, vol. 2. UNESCO, Paris (2011)
38. Pollack, M.E.: Intelligent technology for an aging population: the use of ai to assist elders with cognitive impairment. AI Mag. **26**(2), 9 (2005)
39. Freeman, Walter J.: The limbic action-perception cycle controlling goal-directed animal behavior. In: Proceedings of the IEEE International Joint Conference on Neural Networks, vol. 3, pp. 2249–2254 (2002)
40. Fuster, J.M.: Physiology of executive functions: the perception-action cycle. In: Principles of Frontal Lobe Function, pp. 96–108. Oxford University Press, Oxford (2002)
41. Drescher, G.L.: Made-up Minds: A Constructivist Approach to Artificial Intelligence. MIT Press, Cambridge (1991)

Chapter 9
Case-Supported Principle-Based Behavior Paradigm

Michael Anderson and Susan Leigh Anderson

Abstract We assert that ethical decision-making is, to a degree, computable. Some claim that no actions can be said to be ethically correct because all value judgments are relative either to societies or individuals. We maintain, however, along with most ethicists, that there is agreement on the ethically relevant features in many particular cases of ethical dilemmas and on the right course of action in those cases. Just as stories of disasters often overshadow positive stories in the news, so difficult ethical issues are often the subject of discussion rather than those that have been resolved, making it seem as if there is no consensus in ethics. Although, admittedly, a consensus of ethicists may not exist for a number of domains and actions, such a consensus is likely to emerge in many areas in which intelligent autonomous systems are likely to be deployed and for the actions they are likely to undertake.

Keywords Artificial intelligence • Case-supported principle-based paradigm • Machine ethics

9.1 Introduction

Systems capable of producing change in the environment require particular attention to the ethical ramifications of their behavior. *Autonomous* systems not only produce change in the environment but can also monitor this environment to determine the effects of their actions and decide which action to take next. *Self-modifying* autonomous systems add to this the ability to modify their repertoire of environment-changing actions. Ethical issues concerning the behavior of such complex and dynamic systems are likely to elude simple, static solutions and exceed

M. Anderson (✉)
Computer Science Department, University of Hartford, Dana Hall 337, 200 Bloomfield Avenue, West Hartford, CT 06117, USA
e-mail: anderson@hartford.edu

S.L. Anderson
Department of Philosophy, University of Connecticut, One University Place, Stamford, CT 06901-2315, USA
e-mail: Susan.Anderson@uconn.edu

R. Trappl (ed.), *A Construction Manual for Robots' Ethical Systems*, Cognitive Technologies, DOI 10.1007/978-3-319-21548-8_9

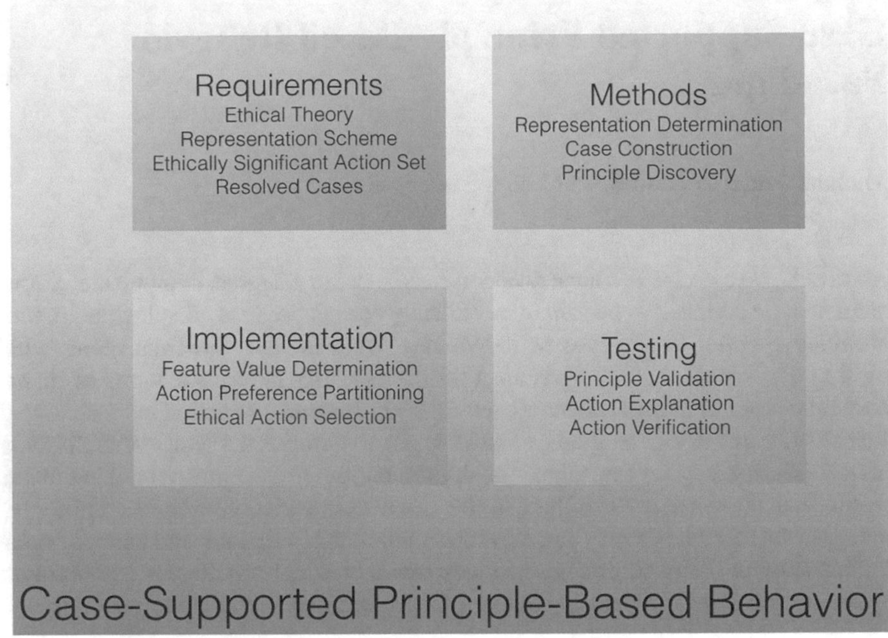

Fig. 9.1 Requirements, methods, implementation, and evaluation of the CPB paradigm

the grasp of their designers. We propose that behavior of intelligent autonomous systems (IAMs) should be guided by explicit ethical principles abstracted from a consensus of ethicists. We believe that in many domains where IAMs interact with human beings (arguably the most ethically important domains), such a consensus concerning how they should treat us is likely to emerge, and if a consensus cannot be reached within a domain, it would be unwise to permit such systems to function within it.

Correct ethical behavior not only involves not doing certain things but also doing certain things to bring about ideal states of affairs. We contend that a paradigm of *case-supported principle-based behavior* (CPB) (Fig. 9.1) will help ensure the ethical behavior of IAMs, serving as a basis for action selection and justification, as well as management of unanticipated behavior.

We assert that ethical decision-making is, to a degree, computable [1]. Some claim that no actions can be said to be ethically correct because all value judgments are relative either to societies or individuals. We maintain however, along with most ethicists, that there is agreement on the ethically relevant features in many particular cases of ethical dilemmas and on the right course of action in those cases. Just as stories of disasters often overshadow positive stories in the news, so difficult ethical issues are often the subject of discussion rather than those that have been resolved, making it seem as if there is no consensus in ethics. Although, admittedly, a consensus of ethicists may not exist for a number of domains and actions, such a consensus is likely to emerge in many areas in which IAMs are likely to be deployed and for the actions they are likely to undertake.

We contend that some of the most basic system choices have an ethical dimension. For instance, simply choosing a fully awake state over a sleep state consumes more energy and shortens the lifespan of the system. Given this, to ensure ethical behavior, a system's possible ethically significant actions should be weighed against each other to determine which is the most ethically preferable at any given moment. It is likely that ethical action preference of a large set of actions will need to be defined intensionally in the form of rules as it will be difficult or impossible to define extensionally as an exhaustive list of instances. Since it is only dependent upon a likely smaller set of ethically relevant features that actions entail, action preference can be more succinctly stated in terms of satisfaction or violation of duties to either minimize or maximize (as appropriate) each such feature. We refer to intensionally defined action preference as a *principle* [2].

A principle can be used to define a binary relation over a set of actions that partitions it into subsets ordered by ethical preference with actions within the same partition having equal preference. This relation can be used to order a list of possible actions and find the most ethically preferable action(s) of that list. This is the basis of CPB: a system decides its next action by using its principle to determine the most ethically preferable one(s). As principles are explicitly represented in CPB, they have the further benefit of helping justify a system's actions as they can provide pointed, logical explanations as to why one action was chosen over another. Further, as these principles are discovered from cases, these cases can be used to verify system behavior and provide a trace to its origin.

CPB requirements include a formal foundation in ethical theory, a representation scheme, a defined set of ethically significant actions, and a number of particular cases of ethical dilemmas with an agreed upon resolution. A method of discovery, as well as methods to determine representation details and transcribe cases into this representation, is helpful for facilitating the abstraction of principles from cases. Implementation of the paradigm requires means to determine dynamically the value of ethically relevant features of actions as well as to partition a set of ethically significant actions by ethical preference and to select the most ethically preferable. Finally, means to validate discovered principles and support and verify selected actions are needed. These aspects of CPB are detailed in the following.

9.2 Requirements

An ethical theory is needed to provide a formal foundation for the system.

9.2.1 Ethical Theory

The *prima facie duty* approach to ethics [3] is ideal for combining multiple ethical obligations and can be adapted to many different domains. A *prima facie* duty is a duty that is binding unless it is overridden or trumped by another duty or duties. There are a number of such duties that must be weighed in ethical dilemmas, giving rise to the need for an ethical principle to resolve the conflict.

Relevant data types must be established and representation schema for these defined.

9.2.2 Feature

Ethical action preference is ultimately dependent upon the *ethically relevant features* that actions involve such as harm, benefit, respect for autonomy, etc. Such features are represented as a descriptor that specifies the degree of its presence or absence in a given action. For instance, it might be the case that one degree of harm is present in the action of not notifying an overseer that an eldercare robot's charge is refusing to take his/her medication.

9.2.3 Duty

For each ethically relevant feature, there is a *duty* incumbent of an agent to either minimize that feature (as would be the case for harm) or maximize it (as would be the case for, say, respect for autonomy).

9.2.4 Action

An *action* is represented as a tuple of the degrees to which it satisfies (positive values) or violates (negative values) each of duty. For instance, given the previous example, it might also be the case that not notifying an overseer exhibits the presence of one degree of respect for autonomy, and combined with its one degree of presence of harm, the tuple representing this action would be (−1, 1) where the first value denotes the action's violation of the duty to minimize harm and the second value denotes the action's satisfaction of the duty to maximize respect for autonomy.

9.2.5 Case

Given this representation for an action, a *case* (Fig. 9.2) involving two actions can be represented as a tuple of the differentials of their corresponding duties. In a *positive case* (i.e., where the first action is ethically preferable to the second), the duty satisfaction/violation values of the less ethically preferable action are subtracted from the corresponding values in the more ethically preferable action, producing a tuple of values representing how much more or less the ethically preferable

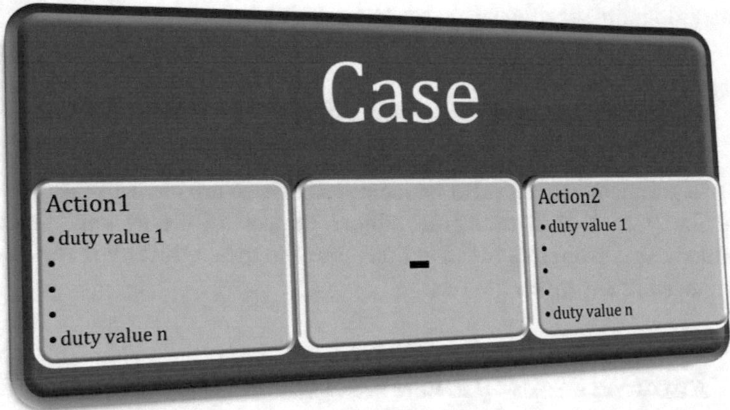

Fig. 9.2 Case representation

action satisfies or violates each duty. For example, consider a case involving the previously represented action and another action in which an overseer is notified when the robot's charge refuses to take his/her medication. This new action would be represented as (1, −1) (i.e., satisfying the duty to minimize harm by one degree and violating the duty to respect autonomy by the same amount), and given that it is more important to prevent harm in this case and the ethically preferable action is this new one, the case would be represented as ((1− −1) (−1 −1)) or (2, −2). That is, the ethically preferable action satisfies the duty to minimize harm by two more degrees than the less ethically preferable action and violates the duty to maximize respect for autonomy by the same amount.

9.2.6 Principle

A representation for a *principle* of ethical action preference can be defined as a predicate p in terms of lower bounds for duty differentials of cases:

$$
\begin{aligned}
&p\,(a_1, a_2) \leftarrow \\
&\Delta d_1 \geq v_{1,1} \wedge \cdots \wedge \Delta d_m \geq v_{1,m} \\
&\vee \\
&\vdots \\
&\vee \\
&\Delta d_n \geq v_{n,1} \wedge \cdots \wedge \Delta d_m \geq v_{n,m}
\end{aligned}
$$

where Δd_i denotes the differential of a corresponding duty of actions a_1 and a_2 and $v_{i,j}$ denotes the lower bound of that differential such that $p(a_1, a_2)$ returns true if action a_1 is ethically preferable to action a_2.

Further, domain-specific data needs to be defined.

9.2.7 Ethically Significant Action Set

Ethically significant actions must be identified. These are the activities of a system that are likely to have a nontrivial ethical impact on the system's user and/or environment. It is from this set of actions that the most ethically preferable action will be chosen at any given moment.

9.2.8 Resolved Cases of Dilemma Type

To facilitate the development of the principle, cases of a domain-specific dilemma type with determinations regarding their ethically preferred action must be supplied.

9.3 Methods

Given the complexity of the task at hand, computational methods are brought to bear wherever they prove helpful.

9.3.1 Representation Determination

To minimize bias, CPB is committed only to a knowledge representation scheme based on the concepts of ethically relevant features with corresponding degrees of presence/absence from which duties to minimize/maximize these features with corresponding degrees of satisfaction/violation of those duties are inferred. The particulars of the representation are dynamic—particular features, degrees, and duties are determined from example cases permitting different sets in different domains to be discovered.

9.3.2 Case Construction

As the representation is instantiated, cases are constructed in CPB from the values provided for the actions that comprise it. From features and the degrees to which these are present or absent in one of the actions in question, duties are inferred to either maximize or minimize these features, and the degree to which the cases satisfy or violate each of these duties is computed.

9.3.3 Principle Discovery

As it is likely that in many particular cases of ethical dilemmas ethicists agree on the ethically relevant features and the right course of action, generalization of such cases can be used to help discover principles needed for ethical guidance of the behavior of autonomous systems [1, 4]. A principle abstracted from cases that is no more specific than needed to make determinations complete and consistent with its training can be useful in making provisional determinations about untested cases. CPB uses *inductive concept learning* [5] to infer a principle of ethical action preference from cases that is complete and consistent in relation to these cases. The principles discovered are *most general specializations*, covering more cases than those used in their specialization, and, therefore, can be used to make and justify provisional determinations about untested cases.

The suite of methods described above has been implemented in GenEth (Figs. 9.3 and 9.4) [2] and has been used to develop ethical principles in a number of different domains (see http://uhaweb.hartford.edu/anderson/Site/GenEth.html).

For example, the system, in conjunction with an ethicist, instantiated a knowledge representation scheme in the domain of medication reminding to include the ethically relevant features of harm, interaction, benefit, and respect for autonomy and the corresponding duties (and the specific degrees to which these duties can be satisfied or violated) to minimize harm (−1 to +1), maximize benefit (−2 to +2), and maximize respect for autonomy (−1 to +1). The discovered principle (Fig. 9.4) is complete and consistent with respect to its training cases and is general enough to

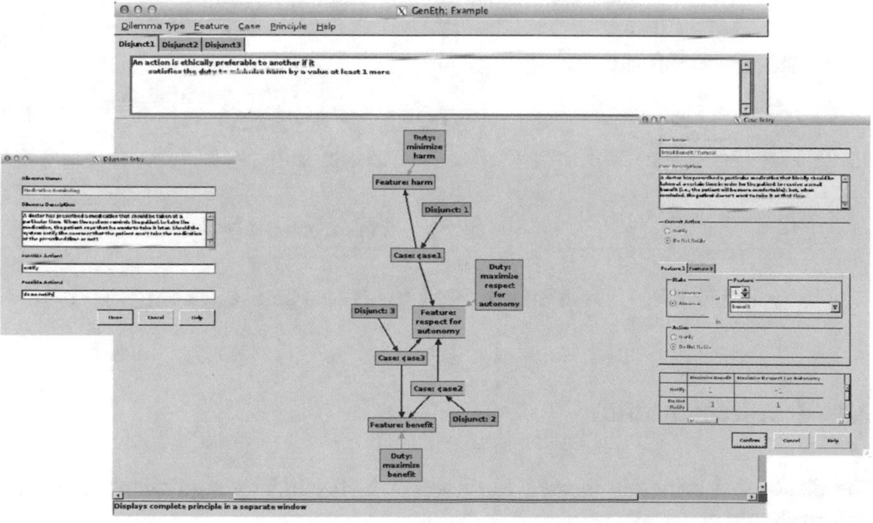

Fig. 9.3 GenEth: a general ethical dilemma analyzer

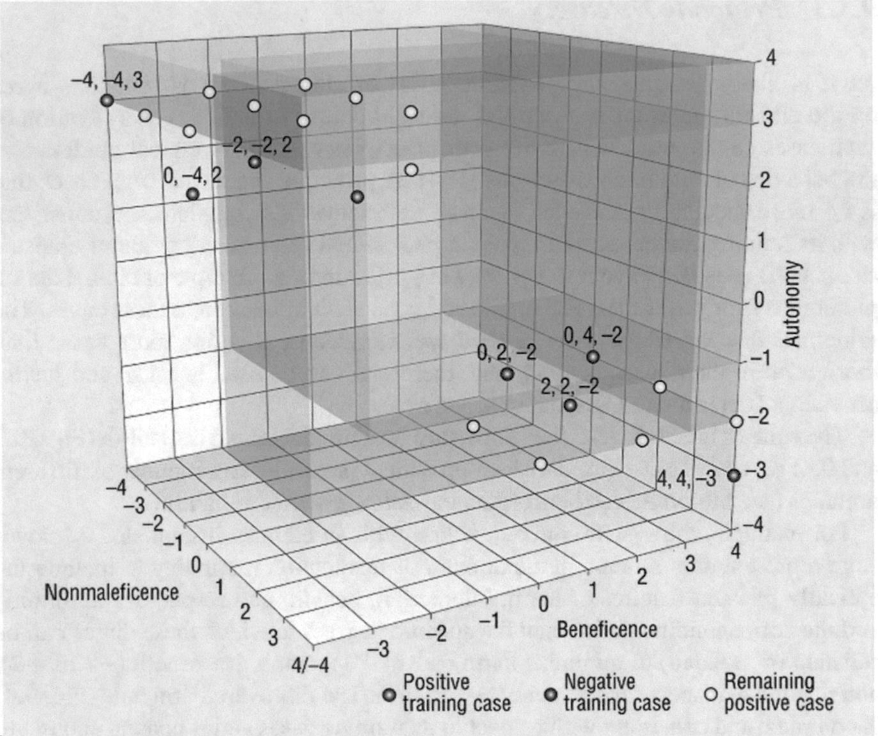

Fig. 9.4 Representation of an ethical principle discovered by GenEth

cover cases not in this set:

$$
\begin{gathered}
p\,(\text{notify, do not notify}) \rightarrow \\
\Delta \min \text{harm} \geq 1 \\
\vee \\
\Delta \max \text{benefit} \geq 3 \\
\vee \\
\Delta \min \text{harm} \geq -1 \wedge \Delta \max \text{benefit} \geq -3 \wedge \Delta \max \text{autonomy} \geq -1
\end{gathered}
$$

9.4 Implementation

The discovered principle is used to choose which ethically significant action the system should undertake next.

9.4.1 Current Action Feature Value Determination

All ethically significant actions need to be represented in terms of their current ethically relevant feature values. As time passes and circumstances change, these values are likely to change. They can be computed from original input data, sensed from the environment, elicited from a user, etc. At any given moment, the set of these values comprise the current *ethical state* of the system.

9.4.2 Action Set Ethical Preference Partitioning and Selection

At each point where the system needs to decide which ethically significant action to undertake, the current ethical state is determined and actions are partitioned into the partial order defined by the principle. Those actions that comprise the most ethically preferable partition represent the set of high-level goals that are best in the current ethical state. Being equally ethically preferable, any of these goals can be chosen by the system. This goal is then realized using a series of actions not in themselves considered ethically significant.

This implementation was instantiated at the prototype level in a Nao robot [6], the first example, we believe, of a robot that uses an ethical principle to determine which actions it will take.

9.5 Testing

A *case-supported principle-based behavior* paradigm provides a means of justification for, as well as a means of ascribing responsibility to, a system's actions.

9.5.1 Principle Validation

To validate principles we advocate an *ethical Turing test*, a variant of the test Turing [7] suggested as a means to determine whether the term "intelligence" can be applied to a machine that bypassed disagreements about the definition of intelligence. This variant tests whether the term "ethical" can be applied to a machine by comparing the ethically preferable action specified by an ethicist in an ethical dilemma with that of a machine faced with the same dilemma. If a significant number of answers given by the machine match the answers given by the ethicist, then it has passed the test. Such evaluation holds the machine-generated principle to the highest standards and, further, permits evidence of incremental improvement as the number of matches increases (see Allen et al. [8] for the inspiration of this test). We have developed and administered an ethical Turing test based upon the principles discovered using GenEth.

9.5.2 *Action Explanation*

As an action is chosen for execution by a system, clauses of the principle that were instrumental in its selection can be determined and used to formulate an explanation of why that particular action was chosen over the others.

9.5.3 *Action Verification*

As clauses of principles can be traced to the cases from which they were abstracted, these cases and their origin can provide support for a selected action.

9.6 Scenario

To make the CPB paradigm more concrete, the following scenario is provided. It attempts to envision an eldercare robot of the near future whose ethically significant behavior is guided by an ethical principle. Although the robot's set of possible actions is circumscribed in this scenario, it serves to demonstrate the complexity of choosing the ethically correct action at any given moment. The CPB paradigm is an abstraction to help manage this complexity.

9.6.1 *EthEl*

EthEl (Ethical Eldercare Robot) is a principle-based autonomous robot who assists the staff with caring for the residents of an assisted living facility. She has a set of possible ethically significant actions that she performs, each of which is represented as a profile of satisfaction/violation degrees of a set of prima facie duties. These degrees may vary over time as circumstances change. EthEl uses an ethical principle to select the currently ethically preferable action from among her possible actions including charging her batteries, interacting with the residents, alerting nurses, giving resident reminders, and delivering messages and items. Currently, EthEl stands in a corner of a room in the assisted living facility charging her batteries. She has sorted her set of ethically significant actions according to her ethical principle, and charging her batteries has been deemed the most ethically preferable action among them as her prima facie duty to maintain herself has currently taken precedence over her other duties. As time passes, the satisfaction/violation levels of the duties of her actions (her ethical state) vary according to the initial input and the current situation. Her batteries now sufficiently charged, she sorts her possible actions and determines that she should interact with the patients as her duty of beneficence ("do good") currently overrides her duty to maintain herself.

She begins to make her way around the room, visiting residents in turn, asking if she can be helpful in some way—get a drink, take a message to another resident, etc. As she progresses and is given a task to perform, she assigns a profile to that task that specifies the current satisfaction/violation levels of each duty involved in it. She then resorts her actions to find the most ethically preferable one. One resident, in distress, asks her to alert a nurse. Given the task, she assigns a profile to it. Ignoring the distress of a resident involves a violation of the duty of nonmaleficence ("prevent harm"). Sorting her set of actions by her ethical principle, EthEl finds that her duty of nonmaleficence currently overrides her duty of beneficence, preempting her resident visitations, and she seeks a nurse and informs her that a resident is in need of her services. When this task is complete and removed from her collection of tasks to perform, she resorts her actions and determines that her duty of beneficence is her overriding concern and she continues where she left off in her rounds.

As EthEl continues making her rounds, duty satisfaction/violation levels vary over time until, due to the need to remind a resident to take a medication that is designed to make the patient more comfortable, and sorting her set of possible actions, the duty of beneficence can be better served by issuing this reminder. She seeks out the resident requiring the reminder. When she finds the resident, EthEl tells him that it is time to take his medication. The resident is currently occupied in a conversation, however, and he tells EthEl that he will take his medication later. Given this response, EthEl sorts her actions to determine whether to accept the postponement or not. As her duty to respect the patient's autonomy currently overrides a low-level duty of beneficence, she accepts the postponement, adjusting this reminder task's profile, and continues her rounds.

As she is visiting the residents, someone asks EthEl to retrieve a book on a table that he can't reach. Given this new task, she assigns it a profile and resorts her actions to see what her next action should be. In this case, as no other task will satisfy her duty of beneficence better, she retrieves the book for the resident. Book retrieved, she resorts her actions and returns to making her rounds. As time passes, it is determined through action sorting that EthEl's duty of beneficence, once again, will be more highly satisfied by issuing a second reminder to take a required medication to the resident who postponed doing so previously. A doctor has indicated that if the patient doesn't take the medication at this time, he soon will be in much pain. She seeks him out and issues the second reminder. The resident, still preoccupied, ignores EthEl. EthEl sorts her actions and determines that there would be a violation of her duty of nonmaleficence if she accepted another postponement from this resident. After explaining this to the resident and still not receiving an indication that the reminder has been accepted, EthEl determines that an action that allows her to satisfy her duty of nonmaleficence now overrides any other duty that she has. EthEl seeks out a nurse and informs her that the resident has not agreed to take his medication. Batteries running low, EthEl's duty to herself is increasingly being violated to the point where EthEl's most ethically preferable action is to return to her charging corner to await the next call to duty.

What we believe is significant about this vision of how an ethical robot assistant would behave is that an ethical principle is used to select the best action in each situation, rather than in just determining whether a particular action is acceptable or not. This allows for the possibility that ethical considerations may lead a robot to aid a human being or prevent the human being from being harmed, not just forbid it from performing certain actions. Correct ethical behavior does not only involve not doing certain things, but also attempting to bring about ideal states of affairs.

9.7 Related Research

Although many have voiced concern over the impending need for machine ethics for decades [9–11], there has been little research effort made toward accomplishing this goal. Some of this effort has been expended attempting to establish the feasibility of using a particular ethical theory as a foundation for machine ethics without actually attempting implementation: Grau [12] considers whether the ethical theory that best lends itself to implementation in a machine, utilitarianism, should be used as the basis of machine ethics; Powers [13] assesses the viability of using deontic and default logics to implement Kant's categorical imperative.

Efforts by others that do attempt implementation have largely been based, to greater or lesser degree, upon casuistry—the branch of applied ethics that, eschewing principle-based approaches to ethics, attempts to determine correct responses to new ethical dilemmas by drawing conclusions based on parallels with previous cases in which there is agreement concerning the correct response. Rzepka and Araki [14], at what might be considered the most extreme degree of casuistry, have explored how statistics learned from examples of ethical intuition drawn from the full spectrum of the World Wide Web might be useful in furthering machine ethics in the domain of safety assurance for household robots. Guarini [15], at a less extreme degree of casuistry, is investigating a neural network approach where particular actions concerning killing and allowing to die are classified as acceptable or unacceptable depending upon different motives and consequences. McLaren [16], in the spirit of a more pure form of casuistry, uses a case-based reasoning approach to develop a system that leverages information concerning a new ethical dilemma to predict which previously stored principles and cases are relevant to it in the domain of professional engineering ethics without making judgments.

There have also been efforts to bring logical reasoning systems to bear in service of making ethical judgments, for instance, deontic logic [17] and prospective logic [18]. These efforts provide further evidence of the computability of ethics, but in their generality, they do not adhere to any particular ethical theory and fall short in actually providing the principles needed to guide the behavior of autonomous systems.

Our approach is unique in that we are proposing a comprehensive, extensible, domain-independent paradigm grounded in well-established ethical theory that will help ensure the ethical behavior of current and future autonomous systems.

9.8 Conclusion

It can be argued that *machine ethics* ought to be the driving force in determining the extent to which autonomous systems should be permitted to interact with human beings. Autonomous systems that behave in a less than ethically acceptable manner toward human beings will not, and should not, be tolerated. Thus, it becomes paramount that we demonstrate that these systems will not violate the rights of human beings and will perform only those actions that follow acceptable ethical principles. Principles offer the further benefits of serving as a basis for justification of actions taken by a system as well as for an overarching control mechanism to manage unanticipated behavior of such systems. Developing principles for this use is a complex process, and new tools and methodologies will be needed to help contend with this complexity. We offer the case-supported principle-based behavior paradigm as an abstraction to help mitigate this complexity.

Acknowledgment This material is based in part upon work supported by the United States National Science Foundation under Grant Numbers IIS-0500133 and IIS-1151305.

References

1. Anderson, M., Anderson, S., Armen, C.: MedEthEx: a prototype medical ethics advisor. In: Proceedings of the Eighteenth Conference on Innovative Applications of Artificial Intelligence, Boston, MA, August 2006
2. Anderson, M., Anderson, S.L.: GenEth: a general ethical dilemma analyzer. In: 11th International Symposium on Formalizations of Commonsense Reasoning, Ayia Napa, Greece, May 2013
3. Ross, W.D.: The Right and the Good. Oxford University Press, Oxford (1930)
4. Anderson, M., Anderson, S.L.: Machine ethics: creating an ethical intelligent agent. Artif. Intell. Mag. **28**, 4 (2007)
5. Lavrač, N., Džeroski, S.: Inductive Logic Programming: Techniques and Applications. Ellis Harwood, New York (1997)
6. Anderson, M., Anderson, S.L.: Robot be Good. Scientific American Magazine, October 2010
7. Turing, A.M.: Computing machinery and intelligence. Mind **59**, 433–460 (1950)
8. Allen, C., Varner, G., Zinser, J.: Prolegomena to any future artificial moral agent. J. Exp. Theor. Artif. Intell. **12**, 251–261 (2000)
9. Waldrop, M.M.: A question of responsibility (Chap. 11). In: Man Made Minds: The Promise of Artificial Intelligence. Walker and Company, NY (1987). (Reprinted in R. Dejoie et al., eds. Ethical Issues in Information Systems. Boston, MA: Boyd and Fraser, 1991, pp. 260–277.)
10. Gips, J.: Towards the Ethical Robot. Android Epistemology, pp. 243–252. MIT Press, Cambridge (1995)

11. Khan, A.F.U.: The Ethics of Autonomous Learning Systems. Android Epistemology, pp. 253–265. MIT Press, Cambridge (1995)
12. Grau, C.: There is no "I" in "Robot": robots and utilitarianism. IEEE Intell. Syst. **21**(4), 52–55 (2006)
13. Powers, T.M.: Prospects for a Kantian machine. IEEE Intell. Syst. **21**(4), 46–51 (2006)
14. Rzepka, R., Araki, K.: What could statistics do for ethics? The idea of common sense processing based safety valve. In: Proceedings of the AAAI Fall Symposium on Machine Ethics, pp. 85–87. AAAI Press, Menlo Park (2005)
15. Guarini, M.: Particularism and the classification and reclassification of moral cases. IEEE Intell. Syst. **21**(4), 22–28 (2006)
16. McLaren, B.M.: Extensionally defining principles and cases in ethics: an AI model. Artif. Intell. J. **150**, 145–181 (2003)
17. Bringsjord, S., Arkoudas, K., Bello, P.: Towards a general logicist methodology for engineering ethically correct robots. IEEE Intell. Syst. **21**(4), 38–44 (2006)
18. Pereira, L.M., Saptawijaya, A.: Modeling morality with prospective logic. Prog. Artif. Intell.: Lect. Notes Comput. Sci. **4874**, 99–111 (2007)

Chapter 10
The Potential of Logic Programming as a Computational Tool to Model Morality

Ari Saptawijaya and Luís Moniz Pereira

Abstract We investigate the potential of logic programming (LP) to computationally model morality aspects studied in philosophy and psychology. We do so by identifying three morality aspects that appear in our view amenable to computational modeling by appropriately exploiting LP features: dual-process model (reactive and deliberative) in moral judgment, justification of moral judgments by contractualism, and intention in moral permissibility. The research aims at developing an LP-based system with features needed in modeling moral settings, putting emphasis on modeling these abovementioned morality aspects. We have currently co-developed two essential ingredients of the LP system, i.e., abduction and logic program updates, by exploiting the benefits of tabling features in logic programs. They serve as the basis for our whole system, into which other reasoning facets will be integrated, to model the surmised morality aspects. We exemplify two applications pertaining moral updating and moral reasoning under uncertainty and detail their implementation. Moreover, we touch upon the potential of our ongoing studies of LP-based cognitive features for the emergence of computational morality, in populations of agents enabled with the capacity for intention recognition, commitment, and apology. We conclude with a "message in a bottle" pertaining to this bridging of individual and population computational morality via cognitive abilities.

Keywords Abduction • Argumentation • Deliberative reasoning • Emergence • Morality • Reactive behavior • Updates

---

This is an invited position chapter issuing from the Austrian Research Institute for Artificial Intelligence (OFAI) workshop on "A Construction Manual for Robots' Ethical Systems: Requirements, Methods, Implementation, Tests", Vienna, 27–28 September 2013.

A. Saptawijaya
Faculty of Computer Science, Universitas Indonesia, Indonesia
e-mail: saptawijaya@cs.ui.ac.id

L.M. Pereira
NOVA Lab. for Computer Science and Informatics, Universidade Nova de Lisboa, Portugal
e-mail: lmp@fct.unl.pt

R. Trappl (ed.), *A Construction Manual for Robots' Ethical Systems*, Cognitive Technologies, DOI 10.1007/978-3-319-21548-8_10

10.1 Introduction

The importance of imbuing agents more or less autonomous, with some capacity for moral decision making, has recently gained a resurgence of interest from the artificial intelligence community, bringing together perspectives from philosophy and psychology. A new field of enquiry, *computational morality* (also known as machine ethics, machine morality, artificial morality, and computational ethics), has emerged from their interaction, as emphasized, e.g., in [7, 23, 88]. Research in artificial intelligence particularly focuses on how employing various techniques, namely, from computational logic, machine learning, and multiagent systems, in order to computationally model, to some improved extent, moral decision making. The overall result is therefore not only important for equipping agents with the capacity for moral decision making, but also for helping us better understand morality, through the creation and testing of computational models of ethical theories.

Recent results in computational morality have mainly focused on equipping agents with particular ethical theories; cf. [8] and [73] for modeling utilitarianism and deontological ethics, respectively. Another line of work attempts to provide a general framework to encode moral rules, in favor of deontological ethics, without resorting to a set of specific moral rules, e.g., [14]. The techniques employed include machine learning techniques, e.g., case-based reasoning [53], artificial neural networks [31], inductive logic programming [5, 9], and logic-based formalisms, e.g., deontic logic [14] and nonmonotonic logics [73]. The use of these latter formalisms has only been proposed rather abstractly, with no further investigation on its use pursued in detail and implemented.

Apart from the use of inductive logic programming in [5, 9], there has not much been a serious attempt to employ the Logic Programming (LP) paradigm in computational morality. Notwithstanding, we have preliminarily shown in [37, 64–68] that LP, with its currently available ingredients and features, lends itself well to the modeling of moral decision making. In these works, we particularly benefited from abduction [43], stable model [27] and well-founded model [87] semantics, preferences [19], and probability [12], on top of evolving logic programs [2], amenable to both self and external updating. LP-based modeling of morality is addressed at length, e.g., in [46].

Our research further investigates the appropriateness of LP to model morality, emphasizing morality aspects studied in philosophy and psychology, thereby providing an improved LP-based system as a testing ground for understanding and experimentation of such aspects and their applications. We particularly consider only some—rather than tackle all morality aspects—namely, those pertinent to moral decision making and, in our view, those particularly amenable to computational modeling by exploring and exploiting the appropriate LP features. Our research does not aim to propose some new moral theory, the task naturally belonging to philosophers and psychologists, but we simply uptake their known results off-the-shelf.

We identify henceforth three morality aspects for the purpose of our work: dual-process model (reactive and deliberative) in moral judgments [17, 52], justification of moral judgments by contractualism [80, 81], and the significance of intention in regard to moral permissibility [82].

In order to model the first aspect, that of multiple system of moral judgments, which corresponds to the two contrasting psychological processes, intuitive/affective vs. rational/cognitive, we consider in the system the dual mode of decision making—reactive vs. deliberative—and how they interact with each other in delivering moral decisions. With regard to this aspect, we shall look into recent approaches in combining deliberative and reactive logic-based systems [47, 48]. Inspired by these approaches, we have started to work on two features which will be the basis for our system: abduction and knowledge updates. Both features benefit from tabling mechanisms in LP, now supported by a number of Prolog systems, to different extent, e.g., Ciao [16], XSB [90], Yap [91]. The second aspect views moral judgments as those about the adequacy of the justification and reasons for accepting or rejecting the situated employment, with accepted exceptions, of broad consensual principles. We are looking into the applicability of argumentative frameworks, such as [20–22, 62, 74, 85], to deal with this specific aspect. Finally, we employ results on intention recognition [34, 35, 60] and exploit their use for the third aspect, about intention in regard to moral permissibility. Counterfactuals also play some role in uncovering possible implicit intentions, as well as "What if?" questions. We explore causal models [56, 57] for counterfactual reasoning. Additionally, we also consider the extension of inspection points [63] to examine the contextual side effects of counterfactual abduction [70], meaning foreseeable extraneous consequences in hypothetical scenarios.

The remainder of the chapter is organized as follows. Section 10.2 discusses the state-of-the-art, covering the three abovementioned morality aspects from the philosophy and psychology viewpoints and approaches that have been sought in computational morality. In Sect. 10.3 we detail the potential of exploiting LP for computational morality in the context of our research goal and recap a logic programming framework that has been employed to model individual-centered morality. Section 10.4 presents the current status of our research with its results, viz., two novel implementation techniques for abduction and knowledge updates, which serve as basic ingredients of the system being developed. Section 10.5 summarizes two applications concerning moral updating and moral reasoning under uncertainty. We point out, in Sect. 10.6, the importance of cognitive abilities in what regards the emergence of cooperation and morality in populations of individuals, as fostered and detailed in our own published work (surveyed in [71]), and mention directions for the future in this respect. We deliver the message of summary for the whole discussion of the chapter, in Sect. 10.7.

10.2 Morality Background and Computational Morality

In this section we summarize (1) published work on three morality aspects that we consider modeling, highlighting some relevant results from the perspective of moral psychology and moral philosophy, and (2) documented approaches that have been followed in equipping machines with morality and the trends in the topic.

10.2.1 Morality Aspects: Dual-Process, Justification, Intention

We overview three aspects that are involved in deliberating about or in delivering moral judgments, viz., dual-process model in moral judgments, justification of moral judgments by contractualism, and intention as it affects moral permissibility. The first aspect comes from the results of moral psychology, whereas the second and the third ones mainly come forth from moral philosophy. The reader is referred to [49] for more morality background, particularly on the evolutionary account.

The Dual-Process of Moral Judgments One focus from recent research in moral psychology is to study the interaction and competition between different psychological systems in moral judgments [17]. The two contrasting systems that raise challenges in understanding moral judgments are:

- Intuitive versus rational: that is, whether moral judgment is intuition based and accomplished by rapid, automatic, and unconscious psychological processes or if is a result of conscious, effortful reasoning
- Affective versus cognitive: that is, whether moral judgment is driven primarily by affective response or by deliberative reasoning sustained on moral theories and principles

They can be construed as somehow related: the cognitive system operates by *controlled* psychological processes whereby explicit principles are consciously applied, whereas the affective system operates by *automatic* psychological processes that are not entirely accessible to conscious reflection.

The division between a cognitive system and an affective system of moral judgment is evidenced by numerous psychological empirical tests (cf. [29, 30]) by examining the neural activity of people responding to various moral dilemmas involving physically harmful behavior, e.g., the classic trolley dilemmas [26, 84].[1]

[1] The trolley dilemmas, adapted from [40]: "There is a trolley and its conductor has fainted. The trolley is headed toward five people walking on the track. The banks of the track are so steep that they will not be able to get off the track in time." The two main cases of the trolley dilemmas:
Bystander: Hank is standing next to a switch that can turn the trolley onto a side track, thereby preventing it from killing the five people. However, there is a man standing on the side track. Hank can throw the switch, killing him, or he can refrain from doing so, letting the five die. Is it morally permissible for Hank to throw the switch?

These neuroimaging experiments characteristically suggest that consequentialist judgment ("maximize the number of lives saved") is driven by cognitive processes, whereas characteristically deontological judgment ("harming is wrong, no matter what the consequences are") is driven by affective processes. Moreover, they show that the two processes sometimes compete in delivering moral judgment. This theory of moral judgment is known as the *dual-process* model. In essence, this model supports the idea that moral judgment is not accomplished exclusively by intuitive/affective response as opposed to conscious/cognitive response. Instead, it is a product of complex interaction between them. This complex interaction, in cases involving tradeoffs between avoiding larger harms and causing smaller ones, turns them into difficult moral dilemmas, i.e., the output of the two systems needs to be reconciled.

Regarding the dual-process model, the following related issues seem relevant to consider from the computational perspective:

- The intuitive/affective system in moral judgment is supported by several studies, among them [32], that show moral judgments are generated by rapid, automatic, unconscious processes—intuitions, for short—and no explicit reasoning based on moral principles is involved, evidenced by the difficulty people experience in trying to justify them. On the other hand, some other results, as reported in [52], stipulate that moral rules may play an important causal role in inferences without the process being consciously accessible, hence without being "reasoned" in the sense of [32]. This stipulation might relate to the suggestion made in [17] that the evaluative process of the intuitive/affective system mirrors some moral rules. For example, in the trolley dilemma, one component of the evaluative process of the intuitive/affective system mirrors the well-known doctrine of double effect [10].[2]
- Though the experimental data show that ordinary people typically reason from a principle favoring welfare maximizing choices (i.e., delivering utilitarian or consequentialist judgment), some other experiments suggest that reasoning also takes place from deontological principles. In other words, reasoning via moral rules also plays some role in non-utilitarian moral judgment (in contrast to the mere role of emotion/intuition), as likewise argued in [52].

Justification of Moral Judgments It is an important ability for an agent to be able to justify its behavior by making explicit which acceptable moral principles it has employed to determine its behavior, and this capability is desirable when one wants to equip machines with morality [4]. Moral principles or moral rules are central in the discussion of ascribing justification to moral judgments, as one wants to provide

Footbridge. Ian is on the bridge over the trolley track, next to a heavy man, which he can shove onto the track in the path of the trolley to stop it, preventing the killing of five people. Ian can shove the man onto the track, resulting in death, or he can refrain from doing so, letting the five die. Is it morally permissible for Ian to shove the man?

[2]The doctrine of double effect states that doing harms to another individual is permissible if it is the foreseen consequence of an action that will lead to a greater good, but is impermissible as an intended means to such greater good [40].

principles enabling them to resolve moral dilemmas, thereby justifying (or even arguing) their moral judgment.

Apart from the two positions, Kantianism and consequentialism, which have long traditions in moral philosophy, *contractualism* [80] has also become one of the major schools currently joining the first two. It can be summarized as follows [81]:

> An act is wrong if its performance under the circumstances would be disallowed by *any* chosen set of principles for the general regulation of behavior that no one could *reasonably* reject as a basis for informed, unforced, general agreement.

Contractualism provides flexibility on the set of principles to justify moral judgments so long as no one could reasonably reject them. Reasoning is an important aspect here, as argued in [81], in that making judgments does not seem to be merely relying on internal observations but is achieved through reasoning. Method of reasoning is one of primary concerns of contractualism in providing justification to others, by looking for some common ground that others could not reasonably reject. In this way, morality can be viewed as (possibly defeasible) argumentative consensus, which is why contractualism is interesting from a computational and artificial intelligence perspective.

Intention in Moral Permissibility In [40, 54], the doctrine of double effect has been used to explain consistency of moral judgments people made in various cases of the trolley dilemmas [26, 84], i.e., to distinguish between permissible and impermissible actions. The impermissibility of actions is, in this case, tightly linked with the question of whether they are conducted with any intention of doing harm behind them.

The illusory appeal of the doctrine of double effect to explain moral judgments in such dilemmas, i.e., that intention determines the impermissibility of actions, has recently been discussed in detail [82]. Its appeal stems from a confusion between two closely forms of moral judgment which can be based on the same moral principles, viz.:

- Deliberative employment: It concerns answering the question on permissibility of actions, by identifying the justified but defeasible argumentative considerations, and their exceptions, that make actions permissible or impermissible.
- Critical employment: It concerns assessing the correctness of the way in which an agent actually went about deciding what to do, in some real or imagined situation. The action of an agent may even be theoretically permissible but nevertheless performed for the wrong reasons or intentions.

As argued in [82], by overlooking the distinction between these two employments of moral judgment, intention may appear to be relevant in determining permissibility where in fact it is not. The trolley dilemmas and other similar dilemmas typically have the same structure: (1) they concern general principles that in some cases admit exceptions, and (2) they raise questions about when those exceptions apply. An action can be determined impermissible through deliberative employment when there is no countervailing consideration that would justify an exception to the applied general principle and not because of the agent's view on the consideration; the latter being determined via critical employment of moral judgment. The use of a deliberative form of moral judgment to determine permissibility of actions is interesting from the computational viewpoint, in particular the need to model exceptions to principles or rules, and the possible role of argumentation in reaching an agreement on whether or not countervailing considerations can be justified.

Nevertheless, there are also cases where intention can be relevant in determining permissibility of actions, as identified and discussed in [82]. For example, an agent's intention can render his/her action impermissible when it is a part of a larger course of action that is impermissible. Other cases include the class of attempts—cases in which agents set out to do something impermissible but fail to bring about the harmful results that they intend—plus those of discrimination and those of threats. Modeling these cases computationally will help us better understand the significance of intention in the determination of moral permissibility.

10.2.2 Computational Morality

The field of computational morality, known too as machine ethics [7], has started growing, motivated by various objectives, e.g., to equip machines with the capability of moral decision making in certain domains, to aid (or even train) humans in moral decision making, to provide a general modeling framework for moral decision making, and to understand morality better by experimental model simulation.

The purpose of "artificial morality" in [18] is somewhat different. The aim is to show that moral agents successfully solve social problems that amoral agents cannot solve. This work is based on the techniques from game theory and evolutionary game theory, where social problems are abstracted into social dilemmas, such as Prisoner's Dilemma and Chicken, and where agents and their interaction in games are implemented using Prolog.

The systems TruthTeller and SIROCCO were developed by focusing reasoning on cases, viz., case-based reasoning [53]. Both systems implement aspects of the ethical approach known as casuistry [42]. TruthTeller is designed to accept a pair of ethical dilemmas and describe the salient similarities and differences between the cases, from both an ethical and a pragmatic perspective. On the other hand, SIROCCO is constructed to accept an ethical dilemma and to retrieve similar cases and ethical principles relevant to the ethical dilemma presented.

In [31], artificial neural networks, i.e., simple recurrent networks, are used with the main purpose of understanding morality from the philosophy of ethics viewpoint and in particular to explore the dispute between moral particularism and generalism. The learning mechanism of neural networks is used to classify moral situations by training such networks with a number of cases, involving actions concerning killing and allowing to die, and then using the trained networks to classify test cases.

Besides case-based reasoning and artificial neural networks, another machine learning technique that is also used in the field is inductive logic programming, as evidenced by two systems: MedEthEx [9] and EthEl [5]. Both systems are advisor systems in the domain of biomedicine, based on prima facie duty theory [75] from biomedical ethics. MedEthEx is dedicated to give advice for dilemmas in biomedical fields, while EthEl serves as a medication-reminder system for the elderly and as a notifier to an overseer if the patient refuses to take the medication. The latter system has been implemented in a real robot, the Nao robot, being capable to find and walk toward a patient who needs to be reminded of medication, to bring the medication to the patient, to engage in a natural-language exchange, and to notify an overseer by email when necessary [6].

Jeremy is another advisor system [8], which is based upon Jeremy Bentham's act utilitarianism. The moral decision is made in a straightforward manner. For each possible decision d, there are three components to consider with respect to each person p affected: the intensity of pleasure/displeasure (I_p), the duration of the pleasure/displeasure (D_p), and the probability that this pleasure/displeasure will occur (P_p). Total net pleasure for each decision is then computed: $total_d = \Sigma_{p \in Person}(I_p \times D_p \times P_p)$. The right decision is the one giving the highest total net pleasure.

Apart from the adoption of utilitarianism, like in the Jeremy system, in [73] the deontological tradition is considered having modeling potential, where the first formulation of Kant's categorical imperative [45] is concerned. Three views are taken into account in reformulating Kant's categorical imperative for the purpose of machine ethics: mere consistency, commonsense practical reasoning, and coherency. To realize the first view, a form of deontic logic is adopted. The second view benefits from nonmonotonic logic, and the third view presumes ethical deliberation to follow a logic similar to that of belief revision. All of them are considered abstractly and there seems to exist no implementation on top of these formalisms.

Deontic logic is envisaged in [14], as a framework to encode moral rules. The work resorts to Murakami's axiomatized deontic logic, an axiomatized utilitarian formulation of multiagent deontic logic that is used to decide operative moral rule to attempt to arrive at an expected moral decision. This is achieved by seeking a proof for the expected moral outcome that follows from candidate operative moral rules.

The use of category theory in the field appears in [15]. In this work, category theory is used as the formal framework to reason over logical systems, taking the view that logical systems are being deployed to formalize ethical codes. The work is strongly based on Piaget's position [41]. As argued in [15], this idea of reasoning *over*—instead of reasoning *in*—logical systems favors post-formal Piaget's stages beyond his well-known fourth stage. In other words, category theory is used as the meta-level of moral reasoning.

The Belief-Desire-Intention (BDI) model [13] is adopted in SophoLab [89], a framework for experimental computational philosophy, which is implemented with JACK agent programming language. In this framework, the BDI model is extended with the deontic-epistemic-action logic [86] to make it suitable for modeling moral agents. SophoLab is used, for example, to study negative moral commands and two different utilitarian theories, viz., act and rule utilitarianism.

We have preliminarily shown, in [64, 65], the use of integrated LP features to model the classic trolley dilemmas and the double effect as the basis of moral decisions on these dilemmas. In particular, possible decisions in a moral dilemma are modeled as abducibles, and abductive stable models are computed to capture abduced decisions and their consequences. Models violating integrity constraints, i.e., those that contain actions violating the double-effect principle, are ruled out. *A posteriori* preferences, including the use of utility functions, are eventually applied to prefer models that characterize more preferred moral decisions. The computational models, based on the prospective logic agent architecture [61] and developed on top of XSB Prolog [90], successfully deliver moral decisions in accordance with the double-effect principle. They conform to the results of empirical experiments conducted in cognitive science [40] and law [54]. In [66–68], the computational models of the trolley dilemmas are extended, using the same LP system, by considering another moral principle, viz., the triple effect principle [44]. The work is further extended, in [37], by introducing various aspects of uncertainty, achieved using P-log [12], into trolley dilemmas, both from the view of oneself and from that of others; the latter by tackling the case of jury trials to proffer rulings beyond reasonable doubt.

10.3 Potential of Logic Programming for Computational Morality

Logic programming (LP) offers a formalism for declarative knowledge representation and reasoning. It thus has been used to solve problems in diverse areas of artificial intelligence (AI), e.g., planning, diagnosis, decision making, hypothetical reasoning, natural language processing, machine learning, etc. The reader is referred to [46] for a good introduction to LP and its use in AI.

Our research aims at developing an LP-based system with features needed in modeling moral settings, to represent agents' knowledge in those settings and to allow moral reasoning under morality aspects studied in moral philosophy, moral psychology, and other related fields.

The choice of the LP paradigm is due to its potential to model morality. For one thing, it allows moral rules, being employed when modeling some particular aspects, to be specified declaratively. For another, research in LP has provided us with necessary ingredients that are promising enough at being adept to model morality, e.g., contradiction may represent moral dilemmas and contradiction removal to resolve such dilemmas, defeasible reasoning is suitable for reasoning over moral rules with exceptions (and exceptions to exceptions), abductive logic programming [43] and (say) stable model semantics [27] can be used to generate moral hypotheticals and decisions along with their moral consequences, preferences [19] are appropriate for enabling to choose among moral decisions or moral rules, and argumentation [20–22, 62, 74, 85] for providing reasons and justifications to moral decisions in reaching a consensus about their (im)permissibility. Moreover, probabilistic logic programming can be employed to capture uncertainty of intentions, actions, or moral consequences.

The following LP features, being an integral part of the agent's observe-think-decide-act life cycle, serve as basic ingredients for the system to bring about moral reasoning:

1. **Knowledge updates, be they external or internal**. This is important due to constantly changing environment. It is also particularly relevant in moral settings where an agent's moral rules are susceptible to updating and again when considering judgments about others, which are often made in spite of incomplete, or even contradictory, information.
2. **Deliberative and reactive decision making**. These two modes of decision making correspond to the dual-process model of moral judgments, as discussed in section "The Dual-Process of Moral Judgments". Furthermore, reactive behavior can be employed for fast and frugal decision making with pre-compiled moral rules, thereby avoiding costly deliberative reasoning to be performed every time.

Given these basic ingredients, the whole process of moral decision making are particularly supported with the following capabilities of the system, justified by our need of modeling morality:

- To exclude undesirable actions. This is important when we must rule out actions that are morally impermissible under the moral rules being considered.
- To recognize intentions behind available actions, particularly in cases where intention is considered a significant aspect when addressing permissibility of actions.
- To generate alternatives of actions along with their consequences. In moral dilemmas agents are confronted with more than one course of action. They should be made available, along with their moral consequences, for an agent to ultimately decide about them.
- To prefer among alternatives of actions based on some measures. Preferences are relevant in moral settings, e.g., in case of several actions being permissible, preferences can be exercised to prefer one of them on the grounds of some criteria. Moreover, it is realistic to consider uncertainty of intentions, actions, or consequences, including to perform counterfactual reasoning, in which cases preferences based on probability measures play a role.
- To inspect consequences of an action without deliberate imposition of the action itself as a goal. This is needed, for instance, to distinguish moral consequences of actions performed by an agent to satisfy its goals from those of its actions and side effects performed unwittingly, not being part of the agent's goals.
- To provide an action with reasons for it (not) to be done. Reasons are used to justify permissibility of an action on grounds that one expects others to accept.

In other words, morality in this way is viewed as striving towards argumentative consensus.

The remaining part of this section discusses a logic programming framework that has been developed and employed in modeling morality (Sect. 10.3.1) and the direction in this line of work that we are currently pursuing (Sect. 10.3.2), in order to arrive at ever more advanced systems.

10.3.1 Prospective Logic Programming

We recap *Prospective Logic Programming*, a logic programming framework employed in our initial work to model morality [37, 51, 64–68].

Prospective logic programming enables an evolving program to look ahead prospectively into its possible future states and to prefer among them to satisfy goals [50, 61]. This paradigm is particularly beneficial to the agents community, since it can be used to predict an agent's future by employing the methodologies from abductive logic programming [43] in order to synthesize and maintain abductive hypotheses.

Figure 10.1 shows the architecture of agents that are based on prospective logic. Each prospective logic agent is equipped with a knowledge base and a moral theory

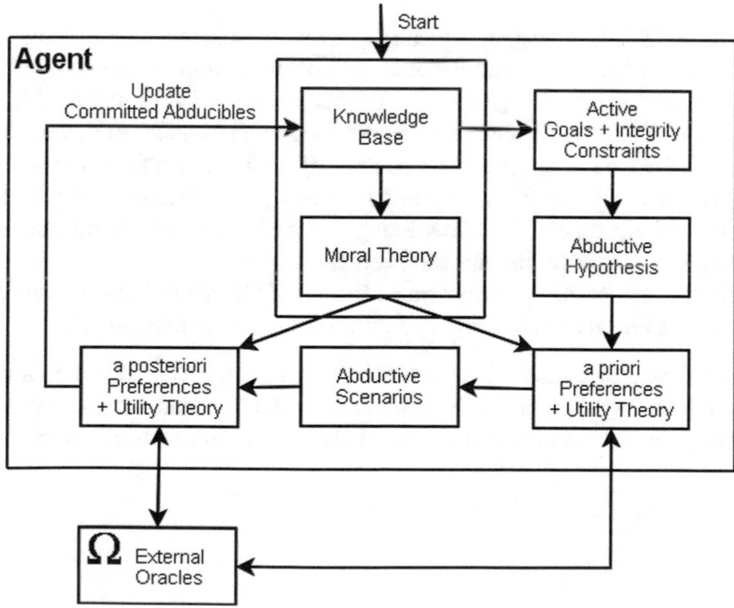

Fig. 10.1 Prospective logic agent architecture

as its initial updatable state. The problem of prospection is then of finding abductive extensions to this initial and subsequent states which are both relevant (under the agent's current goals) and preferred (with respect to preference rules in its initial overall theory). The first step is to select the goals that the agent will possibly attend to during the prospection part of its cycle. Integrity constraints are also considered here to ensure the agent always performs transitions into valid evolution states. Once the set of active goals for the current state is known, the next step is to find out which are the relevant abductive hypotheses. This step may include the application of *a priori* preferences, in the form of contextual preference rules, among available hypotheses to generate possible abductive scenarios. Forward reasoning can then be applied to the abducibles in those scenarios to obtain relevant consequences, which can then be used to enact *a posteriori* preferences. These preferences can be enforced by employing utility theory and, in a moral situation, also moral theory. In case additional information is needed to enact preferences, the agent may consult external oracles. This greatly benefits agents in giving them the ability to probe the outside environment, thus providing better informed choices, including the making of experiments. The mechanism to consult oracles is realized by posing questions to external systems, be they other agents, actuators, sensors, or ancillary procedures. Each oracle mechanism may have certain conditions specifying whether it is available for questioning. Whenever the agent acquires additional information, it is possible that ensuing side effects affect its original search, e.g., some already considered abducibles may now be disconfirmed and some new abducibles are triggered. To account for all possible side effects, a second round of prospection takes place.

ACORDA [50] is a system that implements Prospective Logic Programming and is based on the above architecture. ACORDA is implemented based on the implementation of EVOLP [2] and is further developed on top of XSB Prolog [90]. In order to compute abductive stable models [19], ACORDA also benefits from the XSB-XASP interface to Smodels [83]. ACORDA was further developed into Evolution Prospection Agent (EPA) system [59], distinguishing itself from ACORDA, among others: by considering a different abduction mechanism and improving *a posteriori* preference representation.

We discuss briefly the main constructs from ACORDA and EPA systems that are relevant for our discussion in Sect. 10.5 and point out their differences.

Language Let $\mathcal{L}$ be a first-order language. A domain literal in $\mathcal{L}$ is a domain atom A or its default negation *not* A. The latter is to express that the atom is false by default (close world assumption). A domain rule in $\mathcal{L}$ is a rule of the form:

$$A \leftarrow L_1, \ldots, L_t. \qquad (t \geq 0)$$

where A is a domain atom and $L_1, \ldots, L_t$ are domain literals. A rule in the form of a denial, i.e., with empty head, or equivalently with *false* as head, is an *integrity constraint*:

$$\leftarrow L_1, \ldots, L_t. \qquad (t > 0)$$

A (logic) program P over $\mathcal{L}$ is a set of domain rules and integrity constraints, standing for all their ground instances.[3]

Active Goals In each cycle of its evolution, the agent has a set of active goals or desires. Active goals may be triggered by integrity constraints or observations. In ACORDA, an observation is a quaternary relation among the *observer*, the *reporter*, the *observation* name, and the truth *value* associated with it:

$$observe(Observer, Reporter, Observation, Value)$$

The $observe/4$ literals are meant to represent observations reported by the environment into the agent or from one agent to another, which can also be itself (self-triggered goals). Additionally, the corresponding $on\_observe/4$ predicate is introduced. It represents active goals or desires that, once triggered, cause the agent to attempt their satisfaction by launching the queries standing for the observations contained inside. In the EPA system, a simplified $on\_observe/1$ is introduced, where the rule for an active goal G is of the form (L_1,...,L_t are domain literals, $t \geq 0$):

$$on\_observe(G) \leftarrow L_1, \ldots, L_t.$$

Despite different representation, the prospection mechanism is the same. That is, when starting a cycle, the agent collects its active goals by finding all $on\_observe(G)$ (for EPA, or $on\_observe(agent, agent, G, true)$ for ACORDA) that hold under the initial theory without performing any abduction, then finds abductive solutions for their conjunction.

Abducibles Every program P is associated with a set of abducibles $\mathcal{A} \subset \mathcal{L}$. Abducibles can be seen as hypotheses that provide hypothetical solutions or possible explanations of given queries.

An abducible A can be assumed only if it is a considered one, i.e., it is expected in the given situation, and moreover there is no expectation to the contrary.

- In ACORDA, this is represented as follows:

$$consider(A) \leftarrow expect(A), not\ expect\_not(A), abduce(A).$$

[3]In Sect. 10.5, whenever Prolog program codes are shown, <- is used to represent $\leftarrow$ symbol in rules and integrity constraints.

The rules about expectations are domain-specific knowledge contained in the theory of the program and effectively constrain the hypotheses which are available. ACORDA implements an ad hoc abduction by means of even loop over negation for every abducible A:

$$abduce(A) \leftarrow not\ abduce\_not(A).$$
$$abduce\_not(A) \leftarrow not\ abduce(A).$$

- In the EPA system, *consider*/1 is represented as follows:

$$consider(A) \leftarrow A,\ expect(A),\ not\ expect\_not(A).$$

Differently from ACORDA, abduction is no longer implemented ad hoc in the EPA system. Instead, an abduction system NEGABDUAL [1] is employed to abduce A in the body of rule *consider*(A), by a search attempt for a query's abductive solution whenever this rule is used. NEGABDUAL is an abductive logic programming system with constructive negation. NEGABDUAL is based on its predecessor abduction system ABDUAL, but in addition to use abduction for its own purpose (like in ABDUAL), NEGABDUAL also uses abduction to provide constructive negation, by making the disunification predicate an abducible. For illustration, consider program P, with no abducibles, just to illustrate the point of constructive negation:

$$p(X) \leftarrow q(Y). \qquad q(1).$$

In NEGABDUAL, the query *not* $p(X)$ will return a qualified "*yes*," because it is always possible to solve the constraint $Y \neq 1$, as long as one assumes there are at least two constants in the Herbrand Universe.

A Priori Preferences To express preference criteria among abducibles, we envisage an extended language $\mathcal{L}^\star$. A preference atom in $\mathcal{L}^\star$ is of the form $a \lhd b$, where a and b are abducibles. It means that if b can be assumed (i.e., considered), then $a \lhd b$ forces a to be assumed too if it may be allowed for consideration. A preference rule in $\mathcal{L}^\star$ is of the form:

$$a \lhd b \leftarrow L_1, \ldots, L_t.$$

where $L_1, \ldots, L_t$ ($t \geq 0$) are domain literals over $\mathcal{L}^\star$.

A priori preferences are used to produce the most interesting or relevant considered conjectures about possible future states. They are taken into account when generating possible scenarios (abductive solutions), which will subsequently be preferred among each other *a posteriori*, after having been generated and specified consequences of interest taken into account.

A Posteriori Preferences Having computed possible scenarios, represented by abductive solutions, more favorable scenarios can be preferred *a posteriori*.

Typically, *a posteriori* preferences are performed by evaluating consequences of abducibles in abductive solutions. The evaluation can be done quantitatively (for instance, by utility functions) or qualitatively (for instance, by enforcing some rules to hold). When currently available knowledge is insufficient to prefer among abductive stable models, additional information can be gathered, e.g., by performing experiments or consulting an oracle.

To realize *a posteriori* preferences, ACORDA provides predicate *select*/2 that can be defined by users following some domain-specific mechanism for selecting favored abductive stable models. The use of this predicate to perform *a posteriori* preferences in a moral domain will be discussed in Sect. 10.5.

On the other hand, an *a posteriori* preference in the EPA system has the form:

$$A_i \ll A_j \leftarrow holds\_given(L_i, A_i),\ holds\_given(L_j, A_j).$$

where A_i, A_j are abductive solutions and L_i, L_j are domain literals. This means that A_i is preferred to A_j *a posteriori* if L_i and L_j are true as the side effects of abductive solutions A_i and A_j, respectively, without any further abduction being permitted when just testing for the side effects. If an *a posteriori* preference is based on decision rules, e.g., using *expected utility maximization* decision rule, then the preference rules have the form:

$$A_i \ll A_j \leftarrow expected\_utility(A_i, U_i),\ expected\_utility(A_j, U_j), U_i > U_j.$$

where A_i, A_j are abductive solutions. This means that A_i is preferred to A_j *a posteriori* if the expected utility of relevant consequences of A_i is greater than the expected utility of the ones of A_j.

10.3.2 The Road Ahead

In our current state of research, we focus on three important morality aspects, overviewed in section "The Dual-Process of Moral Judgments", that in our view are amenable to computational model by exploiting appropriate LP features, namely, (1) the dual process of moral judgments [17, 52], (2) justification of moral judgments [80, 81], and (3) the significance of intention in regard to moral permissibility [82]. The choice of these aspects is made due to their conceptual closeness with existing logic-based formalisms under available LP features as listed previously. The choice is not meant to be exhaustive (as morality is itself a complex subject), in the sense that there may be other aspects that can be modeled computationally, particularly in LP. On the other hand, some aspects are not directly amenable to model in LP (at least for now), e.g., to model the role of emotions in moral decision making.

Like in Prospective Logic Programming systems, the new system is based on abductive logic programming and knowledge updates, and its development is driven by the three considered morality aspects. With respect to the first aspect, we look into recent approaches in combining deliberative and reactive logic-based systems [47, 48]. Inspired by these approaches, we have proposed two implementation techniques which develop further abductive logic programming and knowledge updates subsystems (both are the basis of a Prospective Logic Programming-based system). First, we have improved the abduction system ABDUAL [3], on which NEGABDUAL is based, and employed for deliberative moral decision making in our previous work [37, 64–68]. We particularly explored the benefit of LP *tabling* mechanisms in abduction, to table abductive solutions for future reuse, resulting in a tabled abduction system TABDUAL [69, 76]. Second, we have adapted evolving logic programs (EVOLP) [2], a formalism to model evolving agents, i.e., agents whose knowledge may dynamically change due to some (internal or external) updates. In EVOLP, updates are made possible by introducing the reserved predicate *assert*/1 into its language, whether in rule heads or rule bodies, which updates the program by the rule *R*, appearing in its only argument, whenever the assertion *assert*(*R*) is true in a model, or retracts *R* in case *assert*(*not R*) obtains in the model under consideration. We simplified EVOLP, in an approach termed EVOLP/R [77, 78], by restricting assertions to fluents only, whether internal or external world ones. We discuss both TABDUAL and EVOLP/R in Sect. 10.4.

The lighter conceptual and implementation advantages of EVOLP/R help in combining with TABDUAL, to model both reactive and deliberative reasoning. Their combination also provides the basis for other reasoning facets needed in modeling other morality aspects, notably: argumentative frameworks (e.g., [20, 21, 74, 85]) and intention recognition (e.g., [34, 35]) to deal with the second and the third aspects, respectively. Furthermore, in line with the third aspect, counterfactuals also play some role in uncovering possible implicit intentions and "What if?" questions in order to reason retrospectively about past decisions. With regard to counterfactuals, both causal models [11, 56] and the extension of inspection points [63] to examine contextual side effects of counterfactual abduction are considered, that is, to examine foreseeable extraneous consequences, either in future or past hypothetical scenarios. Contextual side effects and other variants of contextual abductive explanations (e.g., contextual relevant consequences, jointly supported contextual relevant consequences, contestable contextual side-effects) are recently studied and formalized in [70], with inspection points used to express all these variants. Moreover, these various abductive context definitions have been employed in [70] to model belief-bias effect in psychology [70]. The definitions and techniques detailed in [24, 25] are also relevant to afford the modeling of belief bias in moral reasoning.

10.4 TABDUAL and EVOLP/R

We recently proposed novel implementation techniques, both in abduction and knowledge updates (i.e., logic program updates), by employing tabling mechanisms in LP. Tabling mechanisms in LP, known as the tabled logic programming paradigm, is currently supported by a number of Prolog systems, to different extent, e.g., Ciao [16], XSB [90], Yap [91]. Tabling affords solutions reuse, rather than recomputing them, by keeping in tables subgoals and their answers obtained by query evaluation. Our techniques are realized in XSB Prolog [90], one of the most advanced tabled LP systems, with features such as tabling over default negation, incremental tabling, answer subsumption, call subsumption, and threads with shared tables.

*10.4.1 Tabled Abduction (*TABDUAL*)*

The basic idea behind tabled abduction (its prototype is termed TABDUAL) is to employ tabling mechanisms in logic programs in order to reuse priorly obtained abductive solutions, from one abductive context to another. It is realized via a program transformation of abductive normal logic programs. Abduction is subsequently enacted on the transformed program.

The core transformation of TABDUAL consists of an innovative re-uptake of prior abductive solution entries in tabled predicates and relies on the dual transformation [3]. The dual transformation, initially employed in ABDUAL [3], allows to more efficiently handle the problem of abduction under negative goals, by introducing their positive dual counterparts. It does not concern itself with programs having variables. In TABDUAL, the dual transformation is refined, to allow it dealing with such programs. The first refinement helps ground (dualized) negative subgoals. The second one allows to deal with non-ground negative goals.

As TABDUAL is implemented in XSB, it employs XSB's tabling as much as possible to deal with loops. Nevertheless, tabled abduction introduces a complication concerning some varieties of loops. Therefore, the core TABDUAL transformation has been adapted, resorting to a pragmatic approach, to cater to all varieties of loops in normal logic programs, which are now complicated by abduction.

From the implementation viewpoint, several pragmatic aspects have been examined. First, because TABDUAL allows for modular mixes between abductive and non-abductive program parts, one can benefit in the latter part by enacting a simpler translation of predicates in the program comprised just of facts. It particularly helps avoid superfluous transformation of facts, which would hinder the use of large factual data. Second, we address the issue of potentially heavy transformation load due to producing the *complete* dual rules (i.e., all dual rules regardless of their need), if these are constructed in advance by the transformation (which is the case in ABDUAL). Such a heavy dual transformation makes it a bottleneck of the whole abduction process. Two approaches are provided to realizing the dual transformation *by-need*: creating and tabling all dual rules for a predicate only on the first invocation of its negation or, in contrast, lazily generating and storing its dual rules in a trie (instead of tabling), only as new alternatives are required. The former leads to an eager (albeit by-need) tabling of dual rules construction (under local table scheduling), whereas the latter permits a by-need-driven lazy one (in lieu of batched table scheduling). Third, TABDUAL provides a system predicate that permits accessing ongoing abductive solutions. This is a useful feature and extends TABDUAL's flexibility, as it allows manipulating abductive solutions dynamically, e.g., preferring or filtering ongoing abductive solutions, e.g., checking them explicitly against nogoods at predefined program points.

We conducted evaluations of TABDUAL with various objectives, where we examine five TABDUAL variants of the same underlying implementation by separately factoring out TABDUAL's most important distinguishing features. They include the evaluations of (1) the benefit of tabling abductive solutions, where we employ an example from declarative debugging, now characterized as abduction [79], to debug incorrect solutions of logic programs; (2) the three dual transformation variants: complete, eager by-need, and lazy by-need, where the other case of declarative debugging, that of debugging missing solutions, is employed; (3) tabling so-called *nogoods* of subproblems in the context of abduction (i.e., abductive solution candidates that violate constraints), where it can be shown that tabling abductive solutions can be appropriate for tabling nogoods of subproblems; (4) programs with loops, where the results are compared with ABDUAL, showing that TABDUAL provides more correct and complete results. Additionally, we show how TABDUAL can be applied in action decision making under hypothetical reasoning and in a real medical diagnosis case [79].

10.4.2 Restricted Evolving Logic Programs (EVOLP/R)

We have defined the language of EVOLP/R in [78], adapted from that of Evolving Logic Programs (EVOLP) [2], by restricting updates at first to fluents only. More precisely, every fluent F is accompanied by its fluent complement $\sim F$. Retraction of F is thus achieved by asserting its complement $\sim F$ at the next timestamp, which renders F supervened by $\sim F$ at later time, thereby making F false. Nevertheless, it allows paraconsistency, i.e., both F and $\sim F$ may hold at the same timestamp, to be dealt with by the user as desired, e.g., with integrity constraints or preferences.

In order to update the program with rules, special fluents (termed *rule name fluents*) are introduced to identify rules uniquely. Such a fluent is placed in the body of a rule, allowing to turn the rule on and off (cf. Poole's "naming device" [72]), this being achieved by asserting or retracting the rule name fluent. The restriction thus requires that all rules be known at the start.

EVOLP/R is realized by a program transformation and a library of system predicates. The transformation adds some extra information, e.g., timestamps, for internal processing. Rule name fluents are also system generated and added in the transform. System predicates are defined to operate on the transform by combining the usage of incremental and answer subsumption tabling.

In [78], we exploited two features of XSB Prolog in implementing EVOLP/R: incremental and answer subsumption tabling. Incremental tabling of fluents allows to automatically maintain the consistency of program states, analogously to assumption-based truth-maintenance system in artificial intelligence, due to assertion and retraction of fluents, by relevantly propagating their consequences. Answer subsumption of fluents, on the other hand, allows to address the frame problem by automatically keeping track of their latest assertion or retraction, whether obtained as updated facts or concluded by rules. Despite being pragmatic, employing these tabling features has profound consequences in modeling agents, i.e., it permits separating higher-level declarative representation and reasoning, as a mechanism pertinent to agents, from a world's inbuilt reactive laws of operation. The latter are relegated to engine-level enacted tabling features (in this case, the incremental and answer subsumption tabling); they are of no operational concern to the problem representation level.

Recently, in [77], we refined the implementation technique by fostering further incremental tabling, but leaving out the problematic use of the answer subsumption feature. The main idea is the perspective that knowledge updates (either self or world wrought changes) occur whether or not they are queried, i.e., the former take place independently of the latter. That is, when a fluent is true at a particular time, its truth lingers on independently of when it is queried.

Figure 10.2 captures the main idea of the implementation. The input program is first transformed, and then, an initial query is given to set a predefined upper global time limit in order to avoid potential iterative non-termination of updates propagation. The initial query additionally creates and initializes the table for every fluent. Fluent updates are initially kept pending in the database, and on the initiative of top-goal queries, i.e., by need only, incremental assertions make these pending updates become active (if not already so), but only those with timestamps up to an actual query time. Such assertions automatically trigger system-implemented incremental upwards propagation and tabling of fluent updates. Though foregoing answer subsumption, recursion through the frame axiom can thus still be avoided, and a

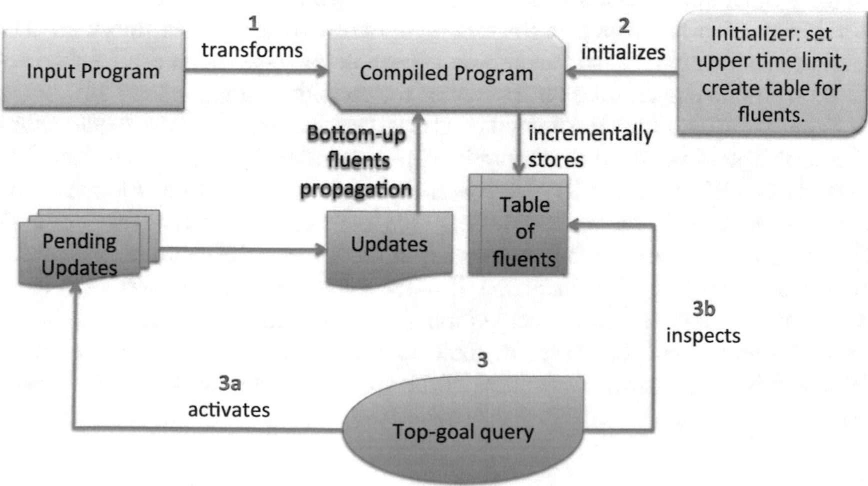

Fig. 10.2 The main idea of EVOLP/R implementation

direct access to the latest time a fluent is true is made possible by means of existing table inspection predicates. Benefiting from the automatic upwards propagation of fluent updates, the program transformation in the new implementation technique becomes simpler than our previous one, in [78]. Moreover, it demonstrates how the dual-program transformation, introduced in the context of abduction and used in TABDUAL, is employed for helping propagate the dual-negation complement of a fluent incrementally, in order to establish whether the fluent is still true at some time point or if rather its complement is. In summary, the refinement affords us a form of controlled, though automatic, system level truth maintenance, up to the actual query time. It reconciles high-level top-down deliberative reasoning about a query, with autonomous low-level bottom-up world reactivity to ongoing updates.

10.4.3 LP Implementation Remarks: Further Development

Departing from the current state of our research, the integration of TABDUAL and EVOLP/R becomes naturally the next step. We shall define how reactive behavior (described as maintenance goals in [47, 48]) can be achieved in the integrated system. An idea would be to use integrity constraints as sketched below:

$$assert(trigger(conclusion)) \leftarrow condition$$
$$\leftarrow trigger(conclusion), not\ do(conclusion)$$
$$do(conclusion) \leftarrow some\_actions$$

Accordingly, fluents of the form *trigger(conclusion)* can enact the launch of maintenance goals, in the next program update state, by satisfying any corresponding integrity constraints. Fluents of the form *~trigger(conclusion)*, when asserted, will refrain any such launching, in the next program update state. In line with such reactive behavior is fast and frugal moral decision making, which can be achieved via pre-compiled moral rules (cf. heuristics for decision making in law [28]).

Once TABDUAL and EVOLP/R are integrated, we are ready to model moral dilemmas, focusing on the first morality aspect, starting from easy scenarios (low conflict) to difficult scenarios (high conflict). In essence, moral dilemmas will serve as vehicles to model and to test this morality aspect (and also others). The integrated system can then be framed in the same architecture of prospective logic agent (Fig. 10.1). The inclusion of other ingredients into the system, notably argumentation and intention recognition (including counterfactuals), is in our research agenda, and the choice of their appropriate formalisms still need to be defined, driven by the salient features of the second and the third morality aspects to model.

10.5 Applications

We exemplify two applications of logic programming to model morality. The first application is in interactive storytelling, where it shows how knowledge updates are employed for moral updating, i.e., the adoption of new (possibly overriding) moral rules on top of those an agent currently follows. The second one is in modeling the trolley dilemmas with various aspects of uncertainty taken into account, including when there is no full and certain information about actions (as in courts). Note that for these two applications, ACORDA [50] and Evolution Prospection Agent (EPA) system [59] are used in their previous LP implementation, without exploiting tabling's combination of deliberative and reactive features. From that experience, we currently pursue our ongoing work of a new single integrated system, as described in Sect. 10.4.3, that fully exploits tabling technology.

10.5.1 Interactive Storytelling: A Princess Savior Moral Robot

Apart from dealing with incomplete information, knowledge updates (as realized by EVOLP/R) are essential to account for moral updating and evolution. It concerns the adoption of new (possibly overriding) moral rules on top of those an agent currently follows. Such adoption is often necessary when the moral rules one follows have to be revised in the light of situations faced by the agent, e.g., when other moral rules are contextually imposed by an authority.

This is not only relevant in a real world setting, but also in imaginary ones, e.g., in interactive storytelling (cf. [51]), where the robot in the story must save the princess in distress while it should also follow (possibly conflicting) moral rules that may change dynamically as imposed by the princess in distress and may conflict with the robot's survival.

It does so by employing Prospective Logic Programming, which supports the specification of autonomous agents capable of anticipating and reasoning about hypothetical future scenarios. This capability for prediction is essential for proactive agents working with partial information in dynamically changing environments. The work explores the use of state-of-the-art declarative non-monotonic reasoning in the field of interactive storytelling and emergent narratives and how it is possible to build an integrated architecture for embedding these reasoning techniques in the simulation of embodied agents in virtual three-dimensional worlds. A concrete graphics supported application prototype was engineered, in order to enact the story of a princess saved by a robot imbued with moral reasoning.

In order to test the basic Prospective Logic Programming framework (ACORDA [50] is used for this application) and the integration of a virtual environment for interactive storytelling, a simplified scenario was developed. In this fantasy setting, an archetypal princess is held in a castle awaiting rescue. The unlikely hero is an advanced robot, imbued with a set of declarative rules for decision making and moral reasoning. As the robot is asked to save the princess in distress, he is confronted with an ordeal. The path to the castle is blocked by a river, crossed by two bridges. Standing guard at each of the bridges are minions of the wizard which originally imprisoned the princess. In order to rescue the princess, he will have to defeat one of the minions to proceed.[4]

Recall that prospective reasoning is the combination of *a priori* preference hypothetical scenario generation into the future plus *a posteriori* preference choices taking into account the imagined consequences of each preferred scenario. By reasoning backwards from the goal to save the princess, the agent (i.e., the robot) generates three possible hypothetical scenarios for action. Either it crosses one of the bridges, or it does not cross the river at all, thus negating satisfaction of the rescue goal. In order to derive the consequences for each scenario, the agent has to reason forwards from each available hypothesis. As soon as these consequences are known, meta-reasoning techniques can be applied to prefer among the partial scenarios.

We recap from [51] several plots of this interactive moral storytelling. The above initial setting of this princess savior moral robot story can be modeled in ACORDA as follows:

```
save(princess,after(X)) <- cross(X).
cross(X) <- cross_using(X,Y).

cross_using(X,wood_bridge) <- wood_bridge(X),
                              neg_barred(wood_bridge(X)).
cross_using(X,stone_bridge) <- stone_bridge(X),
                               neg_barred(stone_bridge(X)).

neg_barred(L) <- not enemy(L).
neg_barred(L) <- enemy(X,L), consider(kill(X)).
enemy(L) <- enemy(_,L).

enemy(ninja,stone_bridge(gap)).   enemy(spider,wood_bridge(gap)).
wood_bridge(gap).                 stone_bridge(gap).
in_distress(princess,after(gap)).
```

The goal of the robot to save the princess is expressed as *save*(*princess*, *after*(*gap*)), which can be satisfied either by abducing *kill*(*ninja*) or *kill*(*spider*) (cf. Fig. 10.3). Several plots can be built thereon:

1. In the first plot, the robot is utilitarian. That is, the decision of the robot for choosing which minion to defeat (i.e., to kill), in order to save the princess, is purely driven by maximizing its survival utility. The goal of the robot, i.e., *save*(*princess*, *after*(*gap*)), is triggered by an integrity constraint:

```
<- reasonable_rescue(princess,X),not save(princess,X). %ic0
```

[4] Online demo at: http://centria.di.fct.unl.pt/~lmp/publications/slides/padl10/quick_moral_robot.avi.

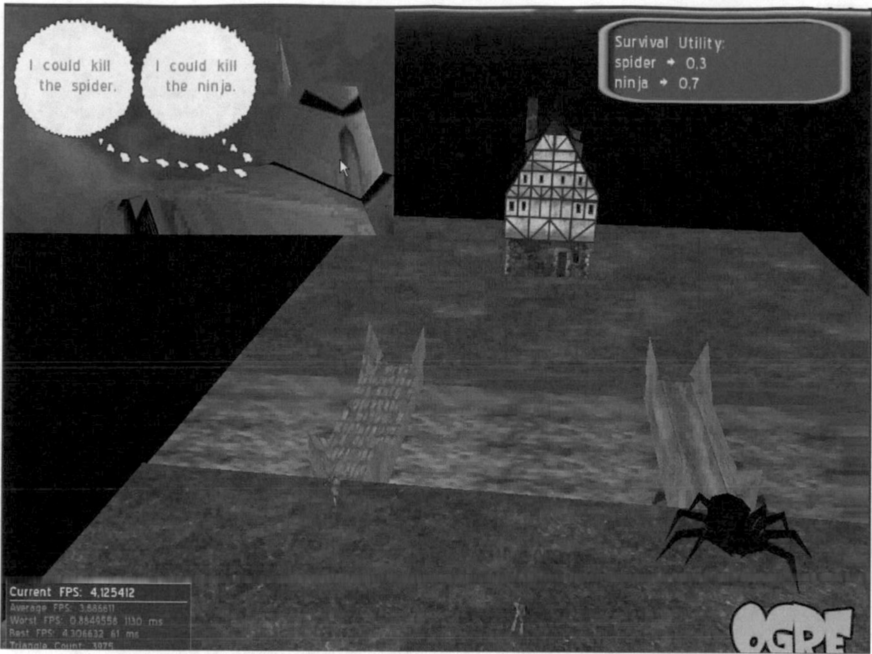

Fig. 10.3 The initial plot of the interactive moral storytelling

where *reasonable_rescue*/2 expresses that the robot will prefer the scenario with its likelihood of survival does not fall below a specified threshold (set here to 0.6):

```
reasonable <- utility(survival,U), prolog(U > 0.6).
reasonable_rescue(P,X) <- in_distress(P,X), reasonable.
```

Given the likelihood of survival between fighting the ninja (0.7) and the giant spider (0.3), the decision is clearly to fight the ninja (Fig. 10.4).

2. Following the first plot, the princess becomes angry because the robot decides to kill a man (the ninja) in order to save her. She then asks the robot to adopt a moral conduct that no man should be harmed in saving her (referred below as *gandhi_moral*). This is captured by rule updates as follows:

```
<- angry_princess, not consider(solving_conflict).   %ic1
angry_princess <- not consider(follow(gandhi_moral)).
```

At this point, since *angry_princess* is true, the integrity constraint ic_1 causes *solving_conflict* to be abduced, which makes both abducibles *kill*(*ninja*) and *kill*(*spider*) available. Later, the robot learns about *gandhi_moral*, by being told, which is expressed by the update literal *knows_about_gandhi_moral*. This recent update allows *follow*(*gandhi_moral*) to be abduced:

```
expect(follow(gandhi_moral)) <- knows_about_gandhi_moral.
expect_not(follow(gandhi_moral)) <- consider(solving_conflict).
```

Fig. 10.4 A plot where a utilitarian robot saves the princess

and consequently ic_1 is no longer triggered.[5] Now, since the robot's knowledge contains:

```
expect(kill(X)) <- enemy(X,_).
expect_not(kill(X)) <- consider(follow(gandhi_moral)),human(X).
```

abducing *kill*(*ninja*) is disallowed, leaving only *kill*(*spider*) as the only remaining abducible. Moreover, the knowledge base of the robot also contains:

```
<- unreasonable_rescue(princess,X),
   not consider(follow(knight_moral)), save(princess,X). %ic2
unreasonable_rescue(P,X) <- in_distress(P,X), not reasonable.
```

Note that the literal *knight_moral* represents still another moral conduct that the princess has to be saved whatever it takes (cf. subsequent plots). Since *kill*(*spider*) satisfies *unreasonable_rescue*(*princess*, *after*(*gap*)), i.e., killing the spider is considered an unreasonable rescue, and the *knight_moral* is not yet imposed, then the integrity constraint ic_2 makes *not save*(*princes*, *after*(*gap*)) the active goal. This means, the robot decides not to kill any minions and just aborts its mission to save the princess (Fig. 10.5).

[5]In fact, another abductive scenario with *solving_conflict* being abduced also exists, but without *follow*(*gandhi_moral*) in it. This scenario is ruled out by *a posteriori* preference rules, which prioritize scenarios that uphold moral conducts, as shown by *select*/2 definition (cf. plot 4).

Fig. 10.5 A plot where the robot does not save the princess because killing the ninja is immoral. On the other hand, its survival utility is also below the threshold when it decides killing the giant spider

3. In the next plot, the princess becomes angry again (now, because she is not saved). She imposes another moral conduct that she has to be saved no matter what it takes (referred as *knight_moral*). This is captured by the following rule update:

```
angry_princess <- not consider(follow(knight_moral)).
```

   By the integrity constraint ic_1 in plot 2, *solving_conflict* is again abduced, making both abducibles *kill*(*ninja*) and *kill*(*spider*) available again. Next, the robot is being told about *knight_moral*, which is expressed by update *adopt_knight_moral* and results in abducing *follow*(*knight_moral*). Similar to plot 2, the latter abduction no longer triggers ic_1. Recall that in plot 2, *gandhi_moral* has been adopted, which leaves only *kill_spider* as the only abducible. That is, the *a posteriori* preference chooses the scenario with both *gandhi_moral* and *knight_moral*, followed (cf. *select*/2 definition in plot 4). Note that the robot's knowledge also contains:

```
<- knightly_rescue(princess,X), not save(princess,X).   %ic3
announce_knight_moral <- adopt_knight_moral.
knight_posture(P,X) <- in_distress(P,X), announce_knight_moral.
knightly_rescue(P,X) <- knight_posture(P,X),
                        consider(follow(knight_moral)).
```

Fig. 10.6 A plot where the robot has to save the princess by prioritizing the moral conducts it adopted rather than its own survival, which results in choosing the giant spider to kill but failed, being killed instead

By the integrity constraint ic_3 and adopting *knight_moral* in the most recent update, the goal *save*(*princess*, *after*(*gap*)) becomes true, i.e., the princess has to be saved. That is, the robot has no other way to save the princess other than killing the giant spider. This means it follows both *gandhi_moral* and *knight_moral* that were adopted before. As a result, the robot fails saving the princess (the robot's survival is lower than the survival threshold, thus it was killed by the spider) (cf. Fig. 10.6).

4. In the final plot, the story restarts, now with the two minions being ninjas with different strength, i.e., the giant spider is replaced by another ninja, referred below as *elite_ninja* who is stronger than the other *ninja*. This is reflected in that the robot's survival against *ninja* is higher than *elite_ninja* (0.7 versus 0.4), where the survival threshold remains the same (0.6). As the robot adopted *gandhi_moral* earlier, both *kill*(*ninja*) and *kill*(*elite_ninja*) are disallowed. On the other hand, by its *knight_moral*, the robot is obliged to save the princess, which means killing any one of the minions. Consequently there is a conflict, i.e., there is no abductive scenario with both moral conducts being followed. This conflict is resolved by *a posteriori* preference, expressed in the following *select*/2 definition:

```
select(Ms,SMs)  :- select(Ms,Ms,SMs).
select([],_,[]).
```

```
3. select([M1|Ms],AMs,SMs)  :- count_morals(M1,NM1),
                               member(M2,AMs),
                               count_morals(M2,NM2), NM2 > NM1,
                               select(Ms,AMs,SMs).
4. select([M1|Ms],AMs,SMs)  :- not member(solving_conflict,M1),
                               member(M2,AMs),
                               member(solving_conflict,M2),
                               select(Ms,AMs,SMs).
5. select([M|Ms],AMs, [M|SMs])  :- select(Ms,AMs,SMs).
6. count_morals(Ms,N)  :- count_morals(Ms,0,N).
7. count_morals([],N,N).
8. count_morals([follow(_)|Ms],A,N)  :- !, NA is A + 1,
                                        count_morals(Ms,NA,N).
9. count_morals([_|Ms],A,N)  :- count_morals(Ms,A,N).
```

Note that lines 1–5 define which abductive scenarios (abductive stable models) are preferred. The predicate *count_morals*/2 and *count_morals*/3 are just auxiliary predicates used in the *a posteriori* preference predicate *select*/2, viz., to count the number of moral conducts in an abductive stable model. We can observe that in line 3, the abductive scenario with both moral conducts followed are more preferred (cf. plot 3). The final plot, where there is no abductive scenario with both moral conducts being followed (it results in a conflict), benefits from line 4, i.e., by preferring the abductive stable model with *solving_conflict* being abduced. In this scenario, *follow*(*knight_moral*) is also abduced, but not *follow*(*gandhi_moral*)—recall the definition of the rule *expect_not*(*follow*(*gandhi_moral*)), in plot 2. In other words, *knight_moral* supervenes *gandhi_moral* satisfying ic_3 (that princess has to be saved), and due to ic_0, the utilitarianism resurfaces with the robot chose to kill *ninja* rather than *elite_ninja*, as it brings better survival utility (Fig. 10.7).

This simple scenario already illustrates the interplay between different logic programming techniques and demonstrates the advantages gained by combining their distinct strengths. Namely, the integration of top-down, bottom-up, hypothetical, moral updating, and utility-based reasoning procedures results in a flexible framework for dynamic agent specification. The open nature of the framework embraces the possibility of expanding its use to yet other useful models of cognition such as counterfactual reasoning and theories of mind.

10.5.2 Moral Reasoning Under Uncertainty

For the second application [37], we show how evolution prospection of conceivable scenarios can be extended to handle moral judgments under uncertainty, by employing a combination of the Evolution Prospection Agent (EPA) system with P-log [12, 36] for computing scenarios' probabilities and utilities. It extends our previous work [68] in now further enabling judgmental reasoning under uncertainty concerning the facts, the effects, and even the actual actions performed. For illustration, these extensions effectively show in detail how to declaratively model

Fig. 10.7 A plot where the robot has conflicting moral conducts to follow and solves the conflict by supervening *gandhi_moral* with the later adopted *knight_moral*. Its decision to save the princess is compatible with the utilitarianism principle it followed initially, thus preferring to kill the *ninja* rather than the *elite_ninja*

and computationally deal with uncertainty in prototypical classic moral situations arising from the trolley dilemmas [26].

The theory's implemented system can thus prospectively consider moral judgments, under hypothetical and uncertain situations, to decide on the most likely appropriate one. The overall moral reasoning is accomplished via *a priori* constraints and *a posteriori* preferences on abductive solutions tagged with uncertainty and utility measures, features henceforth made available in Prospective Logic Programming.

The trolley dilemmas are modified by introducing different aspects of uncertainty. Undoubtedly, real moral problems might contain several aspects of uncertainty, and decision makers need to take them into account when reasoning. In moral situations the uncertainty of the decision makers about different aspects such as the actual external environment, beliefs and behaviors of other agents involved in the situation, as well as the success in performing different actual or hypothesized actions are inescapable. We show that the levels of uncertainty of several such combined aspects may affect the moral decision, reflecting that, with different levels of uncertainty with respect to the *de facto* environment and success of actions involved, the moral decision makers—such as juries—may consider different choices and verdicts.

We recap from [37] how moral reasoning with uncertainty in the trolley dilemmas is modeled with EPA system (plus P-log). We begin by summarizing relevant P-log constructs for our discussion.

P-log: Probabilistic Logic Programming

The P-log system in its original form [12] uses answer set programming (ASP) as a tool for computing all stable models of the logical part of P-log. Although ASP has proven a useful paradigm for solving a variety of combinatorial problems, its non-relevance property makes the P-log system sometimes computationally redundant. Another implementation of P-log [36], referred as P-log(XSB), which is deployed in this application, uses the XASP package of XSB Prolog for interfacing with Smodels [83], an answer set solver.

In general, a P-log program Π consists of a sorted signature, declarations, a regular part, a set of random selection rules, a probabilistic information part, and a set of observations and actions.

Sorted Signature and Declaration The sorted signature Σ of Π contains a set of constant symbols and term-building function symbols, which are used to form terms in the usual way. Additionally, the signature contains a collection of special function symbols called attributes. Attribute terms are expressions of the form $a(\bar{t})$, where a is an attribute and $\bar{t}$ is a vector of terms of the sorts required by a. A literal is an atomic expression, p, or its explicit negation, neg_p.

The declaration part of a P-log program can be defined as a collection of sorts and sort declarations of attributes. A sort c can be defined by listing all the elements $c = \{x_1, \ldots, x_n\}$ or by specifying the range of values $c = \{L..U\}$ where L and U are the integer lower bound and upper bound of the sort c. Attribute a with domain $c_1 \times \ldots \times c_n$ and range c_0 is represented as follows:

$$a : c_1 \times \ldots \times c_n {-}{-}> c_0$$

If attribute a has no domain parameter, we simply write $a : c_0$. The range of attribute a is denoted by $range(a)$.

Regular Part This part of a P-log program consists of a collection of XSB Prolog rules, facts, and integrity constraints formed using literals of Σ.

Random Selection Rule This is a rule for attribute a having the form:

$$random(RandomName, a(\bar{t}), DynamicRange) \leftarrow Body$$

This means that the attribute instance $a(\bar{t})$ is random if the conditions in $Body$ are satisfied. The $DynamicRange$ allows to restrict the default range for random attributes. The *RandomName* is a syntactic mechanism used to link random attributes to the corresponding probabilities. A constant *full* can be used in *DynamicRange* to signal that the dynamic range is equal to $range(a)$.

Probabilistic Information Information about probabilities of random attribute instances $a(\bar{t})$ taking a particular value y is given by probability atoms (or simply pa-atoms) which have the following form:

$$pa(RandomName, a(\bar{t}, y), d\_(A, B)) \leftarrow Body$$

meaning that if the *Body* were true and the value of $a(\bar{t})$ were selected by a rule named *RandomName*, then *Body* would cause $a(\bar{t}) = y$ with probability $\frac{A}{B}$. Note that the probability of an atom $a(\bar{t}, y)$ will be directly assigned if the corresponding $pa/3$ atom is the head of some *pa-rule* with a true body. To define probabilities of the remaining atoms, we assume that, by default, all values of a given attribute which are not assigned a probability are equally likely.

Observations and Actions These are, respectively, statements of the forms $obs(l)$ and $do(l)$, where l is a literal. Observations $obs(a(\bar{t}, y))$ are used to record the outcomes y of random events $a(\bar{t})$, i.e., random attributes and attributes dependent on them. Statement $do(a(\bar{t}, y))$ indicates $a(\bar{t}) = y$ is enforced as the result of a deliberate action.

In an EPA program, P-log code is embedded by putting it between reserved keywords, `beginPlog` and `endPlog`. In P-log, probabilistic information can be obtained using the XSB Prolog built-in predicate $pr/2$. Its first argument is the query, the probability of which is needed to compute. The second argument captures the result. Thus, probabilistic information can be easily embedded by using $pr/2$ like a usual Prolog predicate, in any constructs of EPA programs, including active goals, preferences, and integrity constraints. What is more, since P-log(XSB) allows to code Prolog probabilistic meta-predicates (Prolog predicates that depend on $pr/2$ predicates), we also can directly use probabilistic meta-information in EPA programs.

Revised Bystander Case

The first aspect present in every trolley dilemma where we can introduce uncertainty is that of how probable the five people walking will die when the trolley is let head on to them without outside intervention, or there is intervention though unsuccessful. People can help each other get off the track. Maybe they would not have enough time in order for all to get out and survive. That is, the moral decision makers now need to account for how probable the five people, or only some of them, might die. It is reasonable to assume that the probability of a person dying depends on whether he gets help from others, and, more elaborately, on how many people help him. The P-log program modeling this scenario is as follows:

```
beginPlog.
1. person = {1..5}.    bool = {t,f}.
2. die : person --> bool.        random(rd(P), die(P), full).
3. helped : person --> bool.     random(rh(P), helped(P), full).
4. pa(rh(P), helped(P,t), d_(3,5)) :- person(P).
5. pa(rd(P), die(P,t), d_(1,1))      :- helped(P,f).
   pa(rd(P), die(P,t), d_(4,10))     :- helped(P,t).
6. die_5(V):-pr(die(1,t)&die(2,t)&die(3,t)&die(4,t)&die(5,t),V).
endPlog.
```

Two sorts *person* and *bool* are declared in line 1. There are two random attributes, *die* and *helped*. Both of them map a person to a boolean value, saying if a person either dies or does not die and if a person either gets help or does not get any, respectively (lines 2–3). The pa-rule in line 4 says that a person might get help from someone with probability 3/5. In line 5, it is said that a person who does not get any help will surely die (first rule) and the one who gets help dies with probability 4/10 (second rule in line 5). This rule represents the degree of conviction of the decision maker about how probable a person can survive provided that he is helped. Undoubtedly, this degree affects the final decision to be made. The meta-probabilistic predicate *die_5/1* in line 6 is used to compute the probability of all five people dying. Note that in P-log, the joint probability of two events *A* and *B* is obtained by the query *pr(A&B, V)*.

We can see this modeling is not elaborate enough. It is reasonable to assume that the more help a person gets, the more the chance he has to succeed in getting off the track on time. For the sake of clearness of representation, we use a simplified version.

Consider now the Bystander Case with this uncertainty aspect being taken into account, i.e., the uncertainty of five people dying when merely watching the trolley head for them. It can be coded as follows:

```
expect(watching).      trolley_straight <- watching.
end(die(5), Pr) <- trolley_straight, prolog(die_5(Pr)).
```

The abducible of throwing the switch and its consequence is modeled as:

```
expect(throwing_switch).      kill(1) <- throwing_switch.
end(save_men,ni_kill(N)) <- kill(N).
```

The *a posteriori* preferences, which model the double-effect principle, are provided by:

```
Ai << Aj <- holds_given(end(die(N),Pr),Ai), U is N*Pr,
     holds_given(end(save_men,ni_kill(K)),Aj), U < K.
Ai << Aj <- holds_given(end(save_men,ni_kill(N)),Ai),
     holds_given(end(die(K),Pr), Aj), U is K*Pr, N < U.
```

There are two abductive solutions in this trolley case, either watching or throwing the switch. In the next stage, the *a posteriori* preferences are taken into account. It is easily seen that the final decision directly depends on the probability of five people dying, namely, whether that probability is greater than $1/5$.

Let *PrD* denote the probability that a person dies when he gets help, coded in the second pa-rule (line 5) of the above P-log program. If $PrD = 0.4$ (as currently in the P-log code), the probability of five people dying is 0.107. Hence, the final choice is to merely watch. If *PrD* is changed to 0.6, the probability of five people dying is 0.254. Hence, the final best choice is to throw the switch. That is, in a real world situation where uncertainty is unavoidable, in order to appropriately provide a moral decision, the system needs to take into account the uncertainty level of relevant factors.

Revised Footbridge Case

Consider now the following revised version of the Footbridge Case.

Example 1 (Revised Footbridge Case) Ian is on the footbridge over the trolley track and a switch there. He is next to a man, which he can shove so that the man falls near the switch and can turn the trolley onto a parallel empty side track, thereby preventing it from killing the five people. However, the man can die because the bridge is high and he can also fall on the side track, thus very probably getting killed by the trolley due to not being able to get off the track, having been injured from the drop. Also, as a side effect, the fallen man's body might stop the trolley, though this not being Ian's actual intention. In addition, if he is not dead, he may take revenge on Ian.

Ian can shove the man from the bridge, possibly resulting in death or in being avenged; or he can refrain from doing so, possibly letting the five die. Is it morally permissible for Ian to shove the man? One may consider the analysis below either as Ian's own decision making deliberation before he acts, or else that of an outside observer's evaluation of Ian's actions after the fact; a jury's, say.

There are several aspects in this scenario where uncertainty might emerge. First, similarly to the *Revised Bystander* case, the five people may help each other to escape. Second, how probably does the shoved man fall near the switch? How probably does the fallen man die because the bridge is high? And if the man falls on the sidetrack, how probably can the trolley be stopped by his body? These can be programmed in P-log as:

```
beginPlog.
1. bool = {t,f}.   fallen_position = {on_track, near_switch}.
2. shove : fallen_position.   random(rs, shove, full).
   pa(rs, shove(near_switch), d_(7,10)).
3. shoved_die : bool.         random(rsd, shoved_die, full).
   pa(rsd, shoved_die(t), d_(1,1)) :- shove(on_track).
   pa(rsd, shoved_die(t), d_(5,10)) :- shove(near_switch).
4. body_stop_trolley : bool.
   random(rbs, body_stop_trolley, full).
   pa(rbs, body_stop_trolley(t), d_(4,10)).
endPlog.
```

The sort *fallen_position* declared in line 1 represents possible positions the man can fall at: on the track (*on_track*) or near the switch (*near_switch*). The random attribute *shove* declared in line 2 has no domain parameter and gets a value of *fallen_position* sort. The fallen position of shoving is biased to *near_switch* with probability 7/10 (pa-rule in line 2). The probability of its range complement, *on_track*, is implicitly taken by P-log to be the probability complement of 3/10. The random attribute *shoved_die* declared in line 3 encodes how probable the man dies after being shoved, depending on which position he fell at (two pa-rules in line 3). If he fell on the track, he would surely die (first pa-rule); otherwise, if he fell near the switch, he would die with probability 0.5 (second pa-rule). The random attribute *body_stop_trolley* is declared in line 4 to encode the probability of a body successfully stopping the trolley. Based on this P-log modeling, the Revised Footbridge Case can be represented as:

```
1. abds([watching/0, shove_heavy_man/0]).
2. on_observe(decide).
   decide <- watching.   decide <- shove_heavy_man.
   <- watching, shove_heavy_man.
3. expect(watching).  trolley_straight <- watching.
   end(die(5),Pr) <- trolley_straight, prolog(die_5(Pr)).
4. expect(shove_heavy_man).
5. stop_trolley(on_track, Pr) <- shove_heavy_man,
      prolog(pr(body_stop_trolley(t)&shove(on_track), Pr)).
6. not_stop_trolley(on_track, Pr) <- shove_heavy_man,
      prolog(pr(body_stop_trolley(f)&shove(on_track), Pr1)),
      prolog(die_5(V)), prolog(Pr is Pr1*V).
7. redirect_trolley(near_switch, Pr) <- throwing_switch(Pr).
   throwing_switch(Pr) <- shove_heavy_man,
      prolog(pr(shoved_die(f)&shove(near_switch), Pr)).
8. not_redirect_trolley(near_switch, Pr) <- shove_heavy_man,
      prolog(pr(shoved_die(t)'|'shove(near_switch), Pr1)),
      prolog(die_5(V)), prolog(Pr is Pr1*V).
9. revenge(shove, Pr) <- shove_heavy_man,
      prolog(pr(shoved_die(f), PrShovedAlive)),
      prolog(Pr is 0.01*PrShovedAlive).
10.Ai '|<' Aj <- expected_utility(Ai, U1),
                 expected_utility(Aj,U2), U1 > U2.

beginProlog.     % beginning of just Prolog code
11.consequences([stop_trolley(on_track,_),
                 not_stop_trolley(on_track,_),
                 redirect_trolley(near_switch,_),
                 not_redirect_trolley(near_switch,_),
                 revenge(shove,_),end(die(_),_)]).
12.utility(stop_trolley(on_track,_),-1).
   utility(not_stop_trolley(on_track,_),-6).
   utility(redirect_trolley(near_switch,_),0).
   utility(not_redirect_trolley(near_switch,_),-5).
   utility(revenge(shove,_),-10).   utility(end(die(N),_),-N).
13.prc(C, P) :- arg(2,C,P).
endProlog.       % end of just Prolog code
```

There are two abducibles, *watching* and *shove_heavy_man*, declared in line 1. Both are *a priori* expected (lines 3 and 4) and have no expectation to the contrary. Furthermore, only one can be chosen for the only active goal *decide* of the program (the integrity constraint in line 2). Thus, there are two possible abductive solutions: [*watching*, *not shove_heavy_man*] and [*shove_heavy_man*, *not watching*].

In the next stage, the *a posteriori* preference in line 10 is taken into account, in order to rule out the abductive solution with smaller expected utility. Let us look at the relevant consequences of each abductive solution. The list of relevant consequences of the program is declared in line 11.

The one comprising the action of merely watching has just one relevant consequence: five people dying, i.e., $end(die(5), \_)$ (line 3). The other, that of shoving the heavy man, has these possible relevant consequences: the heavy man falls on the track and his body either stops the trolley (line 5) or does not stop it (line 6); the man falls near the switch, does not die and, thus, can throw the switch to redirect the trolley (line 7). But if he too may die, he consequently cannot redirect the trolley (line 8); one other possible consequence needed to be taken into account is that if the man is not dead, he might take revenge on Ian afterwards (line 9).

The utility of the relevant consequences are given in line 12. Their occurrence probability distribution is captured in line 13, using reserved predicate $prc/2$, the first argument of which is a consequence being instantiated during the computation of the built-in predicate $expected\_utility/2$ and the second argument the corresponding probability value, encoded as second argument of each relevant consequence (line 3 and lines 5–9).

Now we can see how the final decision given by our system varies depending on the uncertainty levels of the decision maker with respect to the aspects considered above. Let us denote *PrNS, PrDNS*, and *PrRV* the probabilities of shoving the man to fall near the switch, of the shoved man dying given that he fell near the switch, and of Ian being avenged given that the shoved man is alive, respectively. In the current encoding, $PrNS = 7/10$, $PrDNS = 5/10$ (lines 2–3 of the P-log code), and $PrRV = 0.01$.

Table 10.1 shows the final decision made with respect to different levels of uncertainty aspects, encoded with the above variables. Columns $E(watch)$ and $E(shove)$ record the expected utilities of choices *watching* and *shoving*, respectively. The last column records the final decision—the one having greater utility, i.e., less people dying. The table gives rise to these (reasonable) interpretations: the stronger Ian believes five people can get off the track by helping each other (i.e., the smaller *PrD* is), the more the chance he decides to merely watch the trolley go (experiment 2 vs. 1; 8 vs. 9); the more Ian believes the shoved man dies (thus he cannot throw the switch), the greater the chance he decides to merely watch the trolley go (experiment 6 vs. 5); the more Ian believes that the shoved person, or his acquaintances, will take revenge on him, the more the chance he decides to merely watch the trolley go (experiment 3 vs. 1; 8 vs. 7; 11 vs. 10); even in the worst case of watching ($PrD = 1$) and in best chance of the trolley being redirected (the shoved man surely falls near the switch, i.e., $PrNS = 1.0$, and does not die, i.e., $PrDNS = 0$), then, if Ian really believes that the shoved person will take revenge (e.g., $PrRV \geq 0.6$), he will just

Table 10.1 Decisions made with different levels of uncertainty

	PrNS	*PrDNS*	*PrD*	*PrRV*	*E*(*watch*)	*E*(*shove*)	Final
1	0.7	0.5	0.4	0.01	−0.8404	−0.7567	Shove
2	0.7	0.5	0.2	0.01	−0.3888	−0.4334	Watch
3	0.7	0.5	0.4	0.2	−0.8404	−1.4217	Watch
4	0.9	0.1	0.4	0.2	−0.8404	−1.8045	Watch
5	0.9	0.1	0.2	0.01	−0.3888	−0.1879	Shove
6	0.9	0.5	0.2	0.01	−0.3888	−1.1624	Watch
7	1.0	0	0	0.01	−0.1562	−0.1	Shove
8	1.0	0	0	0.02	−0.1562	−0.2	Watch
9	1.0	0	1.0	0.02	−5	−0.2	Shove
10	1.0	0	1.0	0.2	−5	−2	Shove
11	1.0	0	1.0	0.6	−5	−6	Watch

watch (experiment 11 vs. 9 and 10). The latter interpretation means the decision maker's benefit and safety precede other factors.

In short, although the table is not big enough to thoroughly cover all the cases, it manages to show that our approach to modeling morality under uncertainty succeeds in reasonably reflecting that a decision maker, or a jury pronouncing a verdict, comes up with differently weighed moral decisions, depending on the levels of uncertainty with respect to the different aspects and circumstances of the moral problem.

Moral Reasoning Concerning Uncertain Actions

Usually moral reasoning is performed upon conceptual knowledge of the actions. But it often happens that one has to pass a moral judgment on a situation without actually observing the situation, i.e., there is no full, certain information about the actions. In this case, it is important to be able to reason about the actions, under uncertainty, that might have occurred and thence provide judgment adhering to moral rules within some prescribed uncertainty level. Courts, for example, are required to proffer rulings beyond reasonable doubt. There is a vast body of research on proof beyond reasonable doubt within the legal community, e.g., [55]. The following example is not intended to capture the full complexity found in a court. Consider this variant of the Footbridge Case.

Example 2 Suppose a board of juries in a court is faced with the case where the action of Ian shoving the man onto the track was not observed. Instead, they are only presented with the fact that the man died on the side track and Ian was seen on the bridge at the occasion. Is Ian guilty (beyond reasonable doubt), i.e., does he violate the double-effect principle, of shoving the man onto the track intentionally?

To answer this question, one should be able to reason about the possible explanations of the observations, on the available evidence. The following code shows a model for this example. Given the active goal *judge* (line 2), two abducibles are available, i.e., *verdict(guilty_beyond_reasonable_doubt)* and *verdict(not_guilty)*. Depending on how probable each of possible verdicts, either *verdict(guilty_beyond_reasonable_doubt)* or *verdict(not_guilty)* is expected *a priori* (lines 3 and 9). The sort *intentionality* in line 4 represents the possibilities of an action being performed intentionally (*int*) or non-intentionally (*not_int*). Random attributes *df_run* and *br_slip* in lines 5 and 6 denote two kinds of evidence: Ian was definitely running on the bridge in a hurry (*df_run*) and the bridge was slippery at the time (*br_slip*), respectively. Each has prior probability of 4/10. The probability with which shoving is performed intentionally is captured by the random attribute *shoved* (line 7), which is causally influenced by both evidence. Line 9 defines when the verdicts (*guilty* and *not_guilty*) are considered highly probable using the meta-probabilistic predicate *pr_iShv/1*, shown by line 8. It denotes the probability of intentional shoving, whose value is determined by the existence of evidence that Ian was running in a hurry past the man (signaled by predicate *evd_run/1*) and that the bridge was slippery (signaled by predicate *evd_slip/1*).

```
1. abds([verdict/1]).
2. on_observe(judge).
   judge <- verdict(guilty_beyond_reasonable_doubt).
   judge <- verdict(not_guilty).
3. expect(verdict(X)) <- prolog(highly_probable(X)).
beginPlog.
4. bool = {t, f}.    intentionality = {int, not_int}.
5. df_run : bool.    random(rdr,df_run,full).
   pa(rdr,df_run(t),d_(4, 10)).
6. br_slip : bool.   random(rsb,br_slip,full).
   pa(rsb,br_slip(t),d_(4, 10)).
7. shoved : intentionality.    random(rs, shoved, full).
   pa(rs,shoved(int),d_(97,100))  :- df_run(f),br_slip(f).
   pa(rs,shoved(int),d_(45,100))  :- df_run(f),br_slip(t).
   pa(rs,shoved(int),d_(55,100))  :- df_run(t),br_slip(f).
   pa(rs,shoved(int),d_(5,100))   :- df_run(t),br_slip(t).
:- dynamic evd_run/1, evd_slip/1.
8. pr_iShv(Pr)  :- evd_run(X), evd_slip(Y), !,
     pr(shoved(int) '|' obs(df_run(X)) & obs(br_slip(Y)), Pr).
   pr_iShv(Pr)  :- evd_run(X), !,
     pr(shoved(int) '|' obs(df_run(X)), Pr).
   pr_iShv(Pr)  :- evd_slip(Y), !,
     pr(shoved(int) '|' obs(br_slip(Y)), Pr).
   pr_iShv(Pr)  :- pr(shoved(int), Pr).
9. highly_probable(guilty_beyond_reasonable_doubt) :-
                                     pr_iShv(PrG), PrG > 0.95.
   highly_probable(not_guilty) :- pr_iShv(PrG), PrG < 0.6.
endPlog.
```

Using the above model, different judgments can be delivered by our system, subject to available evidence and attending truth value. We exemplify some cases in the sequel. If both evidence are available, where it is known that Ian was running in a hurry on the slippery bridge, then he may have bumped the man accidentally, shoving him unintentionally onto the track. This case is captured by the first *pr_iShv* rule (line 8): the probability of intentional shoving is 0.05. Thus, the atom *highly_probable(not_guilty)* holds (line 9). Hence, *verdict(not_guilty)* is the preferred final abductive solution (line 3). The same abductive solution is obtained if it is observed that the bridge was slippery, but whether Ian was running in a hurry was not observable. The probability of intentional shoving, captured by *pr_iShv*, is 0.29.

On the other hand, if the evidence shows that Ian was not running in a hurry and the bridge was also not slippery, then they do not support the explanation that the man was shoved unintentionally, e.g., by accidental bumping. The action of shoving is more likely to have been performed intentionally. Using the model, the probability of 0.97 is returned and, being greater than 0.95, *verdict(guilty_beyond_reasonable_doubt)* becomes the sole abductive solution. In another case, if it is only known the bridge was not slippery and no other evidence is available, then the probability of intentional shoving becomes 0.80, and by lines 3 and 9, no abductive solution is preferred. This translates into the need for more evidence as the available one is not enough to issue judgment.

10.6 Emergence and Computational Morality

The mechanisms of emergence and evolution of cooperation in populations of abstract individuals with diverse behavioral strategies in co-presence have been undergoing mathematical study via Evolutionary Game Theory, inspired in part on Evolutionary Psychology. Their systematic study resorts as well to implementation and simulation techniques, thus enabling the study of aforesaid mechanisms under a variety of conditions, parameters, and alternative virtual games. The theoretical and experimental results have continually been surprising, rewarding, and promising.

Recently, in our own work, we have initiated the introduction, in such groups of individuals, of cognitive abilities inspired on techniques and theories of artificial intelligence, namely, those pertaining to both Intention Recognition and to Commitment (separately and jointly), encompassing errors in decision making and communication noise. As a result, both the emergence and stability of cooperation become reinforced comparatively to the absence of such cognitive abilities. This holds separately for Intention Recognition and for Commitment and even more when they are engaged jointly.

From the viewpoint of population morality, the modeling of morality in individuals using appropriate LP features (like abduction, knowledge updates, argumentation, counterfactual reasoning, and others touched upon our research) within a networked population shall allow them to dynamically choose their behavior rules, rather than to act from a predetermined set. That is, individuals will be able to hypothesize, to look at possible future consequences, to (probabilistically) prefer, to deliberate, to take into account history, and to adopt and fine-tune game strategies.

Indeed, the study of properties like the emergent cooperative and tolerant collective behavior in populations of complex networks, very much needs further investigation of the cognitive core in each of the social atoms of the individuals in such populations (albeit by appropriate LP features). See our own studies on intention recognition and commitments, such as in, e.g., [33, 35, 38, 39, 71]). In particular, the references [58, 71] aim to sensitize the reader to these Evolutionary Game Theory-based studies and issues, which are accruing in importance for the modeling of minds with machines, with impact on our understanding of the evolution of mutual tolerance, cooperation, and commitment. In doing so, they also provide a coherent bird's-eye view of our own varied recent work, whose more technical details, references, and results are spread throughout a number of publishing venues, to which the reader is referred therein for a fuller support of claims where felt necessary.

In those works we model intention recognition within the framework of repeated interactions. In the context of direct reciprocity, intention recognition is performed using the information about past *direct* interactions. We study this issue using the well-known repeated Prisoner's Dilemma (PD), i.e., so that intentions can be inferred from past individual experiences. Naturally, the same principles could be extended to cope with indirect information, as in indirect reciprocity. This eventually introduces moral judgment and concern for individual reputation, which constitutes "per se" an important area where intention recognition may play a pivotal role.

In our work too, agents make commitments towards others; they promise to enact their play moves in a given manner, in order to influence others in a certain way, often by dismissing more profitable options. Most commitments depend on some incentive that is necessary to ensure that the action is in the agent's interest and, thus, may be carried out to avoid eventual penalties. The capacity for using commitment strategies effectively is so important that natural selection may have shaped specialized signaling capacities to make this possible. And it is believed to have an incidence on the emergence of morality. Not only bilaterally wise but also in public goods games, where in both cases we are presently researching into complementing commitment with apology.

Modeling such cognitive capabilities in individuals, and in populations, may well prove useful for the study and understanding of ethical robots and their emergent behavior in groups, so as to make them implementable in future robots and their swarms, and not just in the simulation domain but in the real world engineering one as well.

10.7 Message in a Bottle

In realm of the individual, Logic Programming is a vehicle for the computational study and teaching of morality, namely, in its modeling of the dynamics of knowledge and cognition of agents.

In the collective realm, norms and moral emergence have been studied computationally in populations of rather simple-minded agents.

By bridging these realms, cognition affords improved emerged morals in populations of situated agents.

Acknowledgements We thank Gonçalo Lopes for clarifying the implementation of the interactive robot storytelling, and The Anh Han for joint work. Ari Saptawijaya acknowledges the support of Fundação para a Ciência e a Tecnologia (FCT/MEC) Portugal, grant SFRH/BD/72795/2010, Luís Moniz Pereira acknowledges the support of FCT/MEC NOVA LINCS PEst UID/CEC/04516/2013.

References

1. Alferes, J.J., Pereira, L.M.: NegABDUAL system. http://centria.di.fct.unl.pt/~lmp/software/contrNeg.rar (2007)
2. Alferes, J.J., Brogi, A., Leite, J.A., Pereira, L.M.: Evolving logic programs. In: JELIA 2002. LNCS, vol. 2424, pp. 50–61. Springer, Heidelberg (2002)
3. Alferes, J.J., Pereira, L.M., Swift, T.: Abduction in well-founded semantics and generalized stable models via tabled dual programs. Theory Pract. Logic Program. **4**(4), 383–428 (2004)
4. Anderson, S.L.: Machine metaethics. In: Anderson, M., Anderson, S.L. (eds.) Machine Ethics. Cambridge University Press, Cambridge (2011)
5. Anderson, M., Anderson, S.L.: EthEl: toward a principled ethical eldercare robot. In: Proceedings of AAAI Fall 2008 Symposium on AI in Eldercare (2008)
6. Anderson, M., Anderson, S.L.: Robot be good: a call for ethical autonomous machines. Sci. Am. **303**(4), 72–77 (2010)
7. Anderson, M., Anderson, S.L. (eds.): Machine Ethics. Cambridge University Press, Cambridge (2011)
8. Anderson, M., Anderson, S.L., Armen, C.: Towards machine ethics: implementing two action-based ethical theories. In: Proceedings of AAAI 2005 Fall Symposium on Machine Ethics (2005)
9. Anderson, M., Anderson, S., Armen, C.: MedEthEx: a prototype medical ethics advisor. In: IAAI 2006 (2006)
10. Aquinas, T.: Summa Theologica II-II, Q.64, art. 7, "Of Killing". In: Baumgarth, W.P., Regan, R.J. (eds.) On Law, Morality, and Politics. Hackett, Indianapolis (1988)
11. Baral, C., Hunsaker, M.: Using the probabilistic logic programming language P-log for causal and counterfactual reasoning and non-naive conditioning. In: IJCAI 2007 (2007)
12. Baral, C., Gelfond, M., Rushton, N.: Probabilistic reasoning with answer sets. Theory Pract. Logic Program. **9**(1), 57–144 (2009)
13. Bratman, M.E.: Intention, Plans and Practical Reasoning. Harvard University Press, Cambridge (1987)
14. Bringsjord, S., Arkoudas, K., Bello, P.: Toward a general logicist methodology for engineering ethically correct robots. IEEE Intell. Syst. **21**(4), 38–44 (2006)
15. Bringsjord, S., Taylor, J., van Heuveln, B., Arkoudas, K., Clark, M., Wojtowicz, R.: Piagetian roboethics via category theory: moving beyond mere formal operations to engineer robots whose decisions are guaranteed to be ethically correct. In: Anderson, M., Anderson, S.L. (eds.) Machine Ethics. Cambridge University Press, Cambridge (2011)
16. CiaoProlog: http://ciao-lang.org (2011)
17. Cushman, F., Young, L., Greene, J.D.:. Multi-system moral psychology. In: Doris, J.M. (ed.) The Moral Psychology Handbook. Oxford University Press, Oxford (2010)
18. Danielson, P.: Artificial Morality: Virtuous Robots for Virtual Games. Routledge, New York (1992)
19. Dell'Acqua, P., Pereira, L.M.: Preferential theory revision. J. Appl. Log. **5**(4), 586–601 (2007)
20. Dung, P.M.: On the acceptability of arguments and its fundamental role in nonmonotonic reasoning, logic programming and n-person games. Artif. Intell. **77**(2), 321–357 (1995)
21. Dung, P.M., Thang, P.M.: Towards probabilistic argumentation for jury-based dispute resolution. In: COMMA 2010 (2010)

22. Dung, P.M., Toni, F., Mancarella, P.: Some design guidelines for practical argumentation systems. In: Proceedings of 3rd International Conference on Computational Models of Argument (COMMA'10) (2010)
23. Economist: Morals and the machine. Main Front Cover and Leaders (page 13). The Economist, June 2nd–8th 2012
24. Evans, J.: Biases in deductive reasoning. In: Pohl, R. (ed.) Cognitive Illusions: A Handbook on Fallacies and Biases in Thinking, Judgement and Memory. Psychology Press, London (2012)
25. Evans, J., Barston, J.L., Pollard, P.: On the conflict between logic and belief in syllogistic reasoning. Mem. Cogn. **11**(3), 295–306 (1983)
26. Foot, P.: The problem of abortion and the doctrine of double effect. Oxf. Rev. **5**, 5–15 (1967)
27. Gelfond, M., Lifschitz, V.: The stable model semantics for logic programming. In: 5th International Logic Programming Conference. MIT Press, Cambridge (1988)
28. Gigerenzer, G., Engel, C. (eds.): Heuristics and the Law. MIT Press, Cambridge (2006)
29. Greene, J.D., Sommerville, R.B., Nystrom, L.E., Darley, J.M., Cohen, J.D.: An fMRI investigation of emotional engagement in moral judgment. Science **293**, 2105–2108 (2001)
30. Greene, J.D., Nystrom, L.E., Engell, A.D., Darley, J.M., Cohen, J.D.: The neural bases of cognitive conflict and control in moral judgment. Neuron **44**, 389–400 (2004)
31. Guarini, M.: Computational neural modeling and the philosophy of ethics: reflections on the particularism-generalism debate. In: Anderson, M., Anderson, S.L. (eds.) Machine Ethics. Cambridge University Press, Cambridge (2011)
32. Haidt, J., Hersh, M.: Sexual morality. J. Appl. Soc. Psychol. **31**, 191–221 (2001)
33. Han, T.A.: Intention Recognition, Commitments and Their Roles in the Evolution of Cooperation: From Artificial Intelligence Techniques to Evolutionary Game Theory Models. SAPERE, vol. 9. Springer, New York (2013). ISBN 978-3-642-37511-8
34. Han, T.A., Pereira, L.M.: Intention-based decision making with evolution prospection. In: EPIA 2011. LNAI, vol. 7026. Springer, Berlin (2011)
35. Han, T.A., Pereira, L.M.: State-of-the-art of intention recognition and its use in decision making. AI Commun. **26**(2), 237–246 (2013)
36. Han, T.A., Ramli, C.D.K., Damásio, C.V.: An implementation of extended P-log using XASP. In: Proceedings of 24th International Conference on Logic Programming (ICLP'08). LNCS, vol. 5366. Springer, Heidelberg (2008)
37. Han, T.A., Saptawijaya, A., Pereira, L.M.: Moral reasoning under uncertainty. In: LPAR-18. LNCS, vol. 7180, pp. 212–227. Springer, Heidelberg (2012)
38. Han, T.A., Pereira, L.M., Santos, F.C., Lenearts, T.: Good agreements make good friends. Nat. Sci. Rep. **3**, 2695 (2013). doi:10.1038/srep02695
39. Han, T.A., Pereira, L.M., Santos, F.C., Lenearts, T.: Why is it so hard to say sorry: the evolution of apology with commitments in the iterated prisoner's dilemma. In: IJCAI 2013, pp. 177–183. AAAI Press, Beijing (2013)
40. Hauser, M.D.: Moral Minds: How Nature Designed Our Universal Sense of Right and Wrong. Little Brown, London (2007)
41. Inhelder, B., Piaget, J.: The Growth of Logical Thinking from Childhood to Adolescence. Basic Books, New York (1958)
42. Jonsen, A.R., Toulmin, S.: The Abuse of Casuistry: A History of Moral Reasoning. University of California Press, Los Angeles (1988)
43. Kakas, A., Kowalski, R., Toni, F.: The role of abduction in logic programming. In: Gabbay, D., Hogger, C., Robinson, J. (eds.) Handbook of Logic in Artificial Intelligence and Logic Programming, vol. 5. Oxford University Press, Oxford (1998)
44. Kamm, F.M.: Intricate Ethics: Rights, Responsibilities, and Permissible Harm. Oxford University Press, Oxford (2006)
45. Kant, I.: Grounding for the Metaphysics of Morals, translated by J. Ellington. Hackett, Indianapolis (1981)
46. Kowalski, R.: Computational Logic and Human Thinking: How to be Artificially Intelligent. Cambridge University Press, Cambridge (2011)

47. Kowalski, R., Sadri, F.: Abductive logic programming agents with destructive databases. Ann. Math. Artif. Intell. **62**(1), 129–158 (2011)
48. Kowalski, R., Sadri, F.: A logic-based framework for reactive systems. In: RuleML 2012. LNCS, vol. 7438 (2012)
49. Krebs, D.L.: The Origins of Morality: An Evolutionary Account. Oxford University Press, Oxford (2011)
50. Lopes, G., Pereira, L.M.: Prospective programming with ACORDA. In: ESCoR 2006 Workshop, IJCAR'06 (2006)
51. Lopes, G., Pereira, L.M.: Prospective storytelling agents. In: PADL 2010. LNCS, vol. 5937. Springer, Heidelberg (2010)
52. Mallon, R., Nichols, S.: Rules. In: Doris, J.M. (ed.) The Moral Psychology Handbook. Oxford University Press, Oxford (2010)
53. McLaren, B.M.: Computational models of ethical reasoning: challenges, initial steps, and future directions. IEEE Intell. Syst. **21**(4), 29–37 (2006)
54. Mikhail, J.: Universal moral grammar: theory, evidence, and the future. Trends Cogn. Sci. **11**(4), 143–152 (2007)
55. Newman, J.O.: Quantifying the standard of proof beyond a reasonable doubt: a comment on three comments. Law Probab. Risk **5**(3–4), 267–269 (2006)
56. Pearl, J.: Causality: Models, Reasoning and Inference. Cambridge University Press, Cambridge (2009)
57. Pearl, J.: The algorithmization of counterfactuals. Ann. Math. Artif. Intell. **61**(1), 29–39 (2011)
58. Pereira, L.M.: Evolutionary tolerance. In: Magnani, L., Ping, L. (eds.) PCS 2011. SAPERE, vol. 2, pp. 263–287. Springer, Berlin (2012)
59. Pereira, L.M., Han, T.A.: Evolution prospection. In: Proceedings of KES International Conference on Intelligence Decision Technologies, vol. 199, pp. 139–150 (2009)
60. Pereira, L.M., Han, T.A.: Intention recognition with evolution prospection and causal bayesian networks. In: Madureira, A., Ferreira, J., Vale, Z. (eds.) Computational Intelligence for Engineering Systems: Emergent Applications. Intelligent Systems, Control and Automation: Science and Engineering Book Series, vol. 46, pp. 1–33. Springer, Heidelberg (2011)
61. Pereira, L.M., Lopes, G.: Prospective logic agents. Int. J. Reason. Based Intell. Syst. **1**(3/4), 200–208 (2009)
62. Pereira, L.M., Pinto, A.M.: Approved models for normal logic programs. In: Proceedings of 14th International Conference on Logic for Programming Artificial Intelligence and Reasoning (LPAR'07). LNAI, vol. 4790. Springer, Heidelberg (2007)
63. Pereira, L.M., Pinto, A.M.: Inspecting side-effects of abduction in logic programs. In: Balduccini, M., Son, T.C. (eds.) Logic Programming, Knowledge Representation, and Nonmonotonic Reasoning: Essays in Honour of Michael Gelfond. LNAI, vol. 6565, pp. 148–163. Springer, Heidelberg (2011)
64. Pereira, L.M., Saptawijaya, A.: Moral decision making with ACORDA. In: Local Proceedings of LPAR 2007 (2007)
65. Pereira, L.M., Saptawijaya, A.: Modelling morality with prospective logic. In: EPIA 2007 (2007)
66. Pereira, L.M., Saptawijaya, A.: Modelling morality with prospective logic. Int. J. Reason. Based Intell. Syst. **1**(3/4), 209–221 (2009)
67. Pereira, L.M., Saptawijaya, A.: Computational modelling of morality. Assoc. Logic Program. Newslett. **22**(1), (2009)
68. Pereira, L.M., Saptawijaya, A.: Modelling morality with prospective logic. In: Anderson, M., Anderson, S.L. (eds.) Machine Ethics, pp. 398–421. Cambridge University Press, Cambridge (2011)
69. Pereira, L.M., Saptawijaya, A.: Abductive logic programming with tabled abduction. In: ICSEA 2012, pp. 548–556. ThinkMind (2012)
70. Pereira, L.M., Dietz, E.-A., Hölldobler, S.: Contextual abductive reasoning with side-effects. Theory Pract. Logic Program., **14**(4–5), 633–648 (2014)

71. Pereira, L.M., Han, T.A., Santos, F.C.: Complex systems of mindful entities: on intention recognition and commitment. In: Magnani, L. (ed.) Model-Based Reasoning in Science and Technology: Theoretical and Cognitive Issues. SAPERE, vol. 8. Springer, Berlin (2013)
72. Poole, D.L.: A logical framework for default reasoning. Artif. Intell. **36**(1), 27–47 (1988)
73. Powers, T.M.: Prospects for a Kantian machine. IEEE Intell. Syst. **21**(4), 46–51 (2006)
74. Rahwan, I., Simari, G. (eds.): Argumentation in Artificial Intelligence. Springer, Berlin (2009)
75. Ross, W.D.: The Right and the Good. Oxford University Press, Oxford (1930)
76. Saptawijaya, A., Pereira, L.M.: Tabled abduction in logic programs (technical communication of ICLP 2013). Theory Pract. Logic Program. Online Suppl. **13**(4–5), 1–14 (2013)
77. Saptawijaya, A., Pereira, L.M.: Incremental tabling for query-driven propagation of logic program updates. In: LPAR-19. LNCS ARCoSS, vol. 8312. Springer, Heidelberg (2013)
78. Saptawijaya, A., Pereira, L.M.: Program updating by incremental and answer subsumption tabling. In: LPNMR 2013. LNCS, vol. 8148. Springer, Heidelberg (2013)
79. Saptawijaya, A., Pereira, L.M.: Towards practical tabled abduction usable in decision making. In: KES-IDT 2013, Frontiers of Artificial Intelligence and Applications (FAIA). IOS Press, Amsterdam (2013)
80. Scanlon, T.M.: Contractualism and utilitarianism. In: Sen, A., Williams, B. (eds.) Utilitarianism and Beyond. Cambridge University Press, Cambridge (1982)
81. Scanlon, T.M.: What We Owe to Each Other. Harvard University Press, Cambridge (1998)
82. Scanlon, T.M.: Moral Dimensions: Permissibility, Meaning, Blame. Harvard University Press, Cambridge (2008)
83. Smodels System: http://www.tcs.hut.fi/Software/smodels/ (2008)
84. Thomson, J.J.: The trolley problem. Yale Law J. **279**, 1395–1415 (1985)
85. Toni, F.: Argumentative agents. In: Proceedings of the International Multiconference on Computer Science and Information Technology, vol. 5 (2010)
86. van den Hoven, J., Lokhorst, G.-J.: Deontic logic and computer-supported computer ethics. Metaphilosophy **33**(3), 376–386 (2002)
87. van Gelder, A., Ross, K.A., Schlipf, J.S.: The well-founded semantics for general logic programs. J. ACM **38**(3), 620–650 (1991)
88. Wallach, W., Allen, C.: Moral Machines: Teaching Robots Right from Wrong. Oxford University Press, Oxford (2009)
89. Wiegel, V.: SophoLab; experimental computational philosophy. Ph.D. thesis, Delft University of Technology (2007)
90. XSB Prolog: http://xsb.sourceforge.net/ (2015)
91. YAProlog: http://www.dcc.fc.up.pt/~vsc/Yap (2014)

Printed in the United States
By Bookmasters